DEUTSCHE FORSCHUNGSANSTALT FÜR
LUFT- UND RAUMFAHRT E.V. (DLR)

SOLAR THERMAL CENTRAL RECEIVER SYSTEMS

Volume 3:
Performance Evaluation Standards for Solar Central Receivers

Editors: M. Carasso and M. Becker

Springer-Verlag Berlin Heidelberg GmbH 1990

Meir Carasso, Ph. D.
Solar Energy Research Institute (SERI), Golden, Colorado, USA

Dr.-Ing. Manfred Becker,
Deutsche Forschungsanstalt für Luft- und Raumfahrt e.V. (DLR),
Hauptabteilung Energietechnik, Köln

ISBN 978-3-540-53270-5   ISBN 978-3-642-88196-1 (eBook)
DOI 10.1007/978-3-642-88196-1

2362/3020−543210

Preface and Acknowledgements

The need for international standards for evaluating the performance of solar thermal receivers was first recognized formally at the 3rd International Workshop on Solar Central Receiver Systems held in Konstanz, Germany, in 1986. At the ensuing IEA/SSPS Task III (Receiver Technology) meeting, A. Baker of Sandia National Laboratories in Livermore, California, suggested an approach. Subsequently, the Solar Energy Research Institute (SERI) was assigned the responsibility of organizing an international task force to produce a document that would meet the needs of the international solar thermal community and reflect agreements on definitions of nomenclature, performance parameters, and methods for evaluating them. This report is the culmination of the work of that task force.

We thank the principal authors who were responsible for individual chapters and all those who contributed to the chapters, as well as W. Durisch of the Paul Scherrer Institute of Switzerland for his early involvement in the task force. The editors wish to thank all the reviewers, from both government and private organizations in Europe and the USA, for their valuable comments. Thanks are also due for diligent typing and editorial support to J. Fried, K. Vernon, D. Sayler, P. Haefele, F. VanDerPol, I. Medina, and P. Pitchford of SERI, USA, and to U. Rachow at DLR in Germany.

The editors wish to express gratitude for the opportunity of being a part of this international cooperative effort and in this way making a contribution toward improving international understanding.

This work was supported in part by the U.S. Department of Energy's Solar Thermal Program and in part by the solar programs in the German Federal Ministry for Research and Technology and in the Spanish Ministry for Industry and Energy.

M. Carasso, USA
M. Becker, Germany

Contents

Introduction

The thermal receiver is a key component of any solar thermal central receiver plant. It is located at the focal point of the solar heliostats, and its function is typically to convert the incoming solar flux to thermal energy in a working fluid or, in receiver/reactors, to chemical energy in the reactants. The efficiency with which a receiver accomplishes this function is of central interest in solar thermal technology. The higher the efficiency, the lower the cost of energy produced. Over the last decade, many different kinds of receiver concepts have been built. The performance of these receivers is first predicted on the basis of the design, and then it is evaluated after a prototype has been built.

Solar thermal receivers have been built and operated in many countries. Researchers and technology specialists have evaluated the performance of these receivers using many different standards and definitions, as might be expected, to suit different needs and to reflect different approaches and methods. These experiences have provided a rich knowledge base on receiver performance and its evaluation. At the same time, the variety of standards and methods used by different researchers has caused occasional difficulties in the interpretation of performance evaluation results reported in the literature.

The motivation for this report, therefore, derives from two sources. First is the need to facilitate the evaluation of the performance of solar receivers by providing a central document defining standards for such undertakings; second, to enable an easy interpretation and a comparative evaluation of different receiver concepts for which performance has been evaluated in different countries.

This report is concerned exclusively with the task of evaluating the performance of an existing receiver, and not, for example, with predicting the performance of a receiver concept or design before it has been built. Moreover, whenever a system component--in this case, the receiver--is excluded from the whole system (which typically includes, in addition, a heliostat field and a conversion system of some kind) for the purpose of performance evaluation, some risk is involved in selecting the assumptions that inevitably have to be made. Aware of these pitfalls, the reader is cautioned to examine the definitions used and to recall that this document is not specifically concerned either with receiver design, on the one hand, nor with receiver or system optimization, on the other.

Following the same line of reasoning, the concepts of essergy and thermodynamic availability have not been used in the evaluation standards recommended in this volume. They are most useful as system design concepts rather than receiver performance parameters, and, more practically, they have very seldom been used as receiver performance evaluation criteria by the solar thermal community.

In selecting performance evaluation parameters, the primary emphasis is on the measurement of physical quantities that contain information that can be used to evaluate different aspects that make up "receiver performance." To provide a common reference point for defining the parameters as well as the measurement and calculation procedures that constitute these standards, it has been necessary to begin by developing a common language consisting of general definitions, a nomenclature, and specific definitions of terms useful in describing the thermodynamic processes of solar thermal receivers. These make up the first two chapters of this report. Chapter 3 deals specifically with defining the small set of parameters that are necessary for both the technical and economic evaluation of the performance of solar thermal receivers. Chapters 4 through 7 describe the methods that have been found in the experience of the solar thermal community to be the most useful and accurate in measuring these parameters or otherwise estimating them. This structure is depicted in the diagram below.

Overview of the Structure of this Report

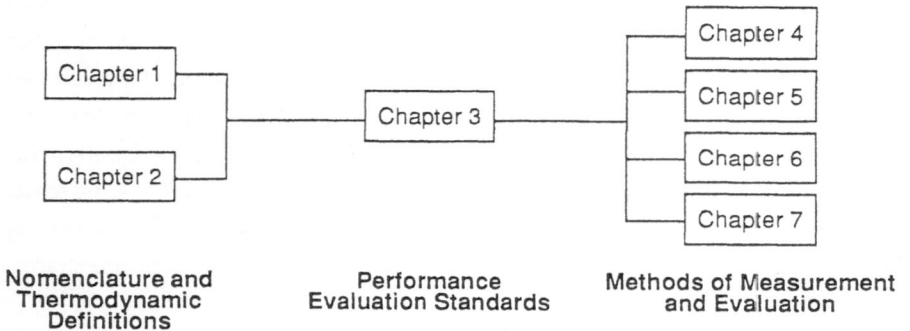

| Nomenclature and Thermodynamic Definitions | Performance Evaluation Standards | Methods of Measurement and Evaluation |

As might be expected, and will become clear to the reader, a number of approaches can be taken to evaluate a receiver's performance. Professionals attempting to evaluate the performance of receivers are encouraged, whenever possible, to use more than one method of estimating key parameters and to compare and report the results they obtain using different methods. An evaluation of the error involved in these estimates greatly enhances the quality of reported results.

Individual chapters of this report were contributed by authors from the Federal Republic of Germany, Spain, and the U.S.A. The names of the authors appear at the beginning of each chapter. Nonetheless, considerable effort has gone into making the document self-consistent, complete, and nonduplicative. Many examples are given to illustrate measurement and calculation methods. For those interested in studying particular aspects in a given chapter in greater detail, a list of references has been included at the end of each chapter.

A balance between brevity and completeness has been attempted by assuming that the user of this report has a background in mechanical engineering. Thus, the discussion focuses rather rapidly on aspects that are specific to solar thermal applications and avoids general introductory material; on the other hand, the discussion does not purport to contain an exhaustive treatment of many of the disciplines involved (e.g., heat transfer, fluid mechanics, or thermodynamics).

The authors hope that this report will prove to be useful. If so, the document will undoubtedly be updated as new receiver concepts, measurement methods, and performance definitions emerge.

M. Carasso
July 1990

Chapter 1

General Definitions and Nomenclature

General Definitions and Nomenclature

M. Sánchez
CIEMAT-IER, Spain

With contributions by
C. Gómez Camacho, University of Seville, Spain

Contents

1.1 Solar Receiver Systems

In a solar central receiver power plant, the receiver system converts concentrated radiant energy reflected from the heliostat field into thermal energy
contained in a heat transport fluid. To date, all pilot-plant designs have
been of the tube receiver type; that is, the heat transport fluid flows
through tubes, usually arranged in multiple panels. These panels are absorbing surfaces similar to conventional fossil-fuel boilers.

Recent research and development activities are being carried out on other concepts in which the fluid does not flow in tubes (e.g., volumetric or direct-
absorption receivers). However, no extensive tests on a pilot scale (1 MW_t)
have yet been reported.

1.1.1 Receiver System Components

The main components of the receiver system are as follows:

• Absorber

• Interconnecting pipes, headers, tanks, etc.

• Receiver structure required to support panels, interconnecting pipes, tanks,
  insulating material, doors, etc.

• Insulating material used both as insulation for panels and pipes and as
  thermal shielding of the tower or structure

• Instrumentation and controls necessary to maintain the receiver heat transport fluid conditions and to protect the receiver from emergency conditions.

1.1.2 Design Options

Two primary aspects of a receiver are the configuration and the working fluid.

1. Configuration. There are two general receiver configurations: cavity and
   external. In a cavity receiver, radiation reflected from the heliostat
   field passes through an aperture into a box-like structure (cavity).

Heat-transfer surfaces are located inside the cavity. A part of the radiant flux that reaches an element of the absorber surface is reflected. All surfaces emit radiation, some of which is lost to the environment.

In external receivers, heat-absorbing surfaces are exposed to the heliostat field. Radiant flux that reaches the absorber surface originates entirely from the heliostat field. For large power plants, the external receiver is typically a multipanel polyhedron that approximates a cylinder. For small plants, a flat panel that faces north--the so-called billboard type--is used.

From the perspective of receiver performance, external and cavity receivers are different in several ways. The emissivity losses of external receivers are generally higher than those of cavity receivers because the view factors of external receiver panels are larger. For this reason, reflection losses are also slightly higher for external receivers. Spillage losses (radiative flux that "spills" over the absorbing surfaces), however, are generally higher for cavity receivers. This is due to their small targets (aperture planes). Convection losses may also be higher in cavity receivers because the heated surface area is large, i.e., it includes the active as well as the inactive (floor, roof, etc.) areas of the cavity.

2. <u>Heat transfer fluid</u>. Several types of working fluids (the term "working fluid" is interchangeable with the term "heat transfer fluid") are used in central receivers; to date, water/steam, molten salt, sodium, and air have been used. Because water/steam systems require only a single fluid for the receiver and the turbine, direct coupling is possible.

Molten salt and sodium are both single-phase fluids working at low pressures. Both display superior thermal conductivity properties, resulting in higher allowable peak fluxes and allowing thinner tubes to be used. However, both molten salt and sodium present difficulties in that they freeze at temperatures well above ambient, thus requiring trace heating to maintain a liquid state.

Sodium has better heat transfer characteristics than molten salt, but salt
has a higher heat capacity, which is important for storage. Sodium also
presents some important safety problems, because of its high reactivity
with water (presenting the possibility of explosion) and air (presenting
the possibility of fire). The advantage of air receivers is that they can
operate at high temperatures and are restricted only by materials.

1.2 Classification and Identification of Solar Receivers

The receiver is the unit that converts concentrated solar radiation into heat
or chemical energy. Solar receivers can be classified according to their
energy-exchange processes. Table 1-1 describes the main energy-exchange process of solar receivers in high-temperature applications.

1.2.1 Indirect-Energy-Exchange Receivers

Indirect-energy-exchange receivers are those in which concentrated solar radiation reaches an absorber surface that transfers energy to a heat transfer
fluid. Such an absorber may be metallic or ceramic, depending on the temperature range in which the receiver is designed to operate. There are two groups
of indirect-energy-exchange receivers:

1. Tube receivers. The absorber surface is composed of tube panels, in which
    a heat transfer fluid flows in parallel through each tube of the same
    panel. Depending on the design, panels may be connected in parallel or in
    series. Depending on the position of the absorber surface, tube receivers
    are also classified as cavity and external receivers.

    Cavity receivers allow concentrated solar radiation through an aperture.
    Inside the cavity, the flux diverges and reaches the heat-exchange panels.
    In contrast, concentrated solar radiation impinges directly onto external
    receiver panels.

Table 1-1. Energy-Exchange Processes of Solar Receivers

Type	Energy Exchange	Receiver Groups	High-Temperature Solar Applications
Indirect Energy Exchange	Transfer energy to an absorber surface	Tube receivers	Solar energy to heat transfer fluid via conduction and convection; chemical reactions
		Volumetric receivers	Solar energy to heat transfer fluid by convection; chemical reactions
Direct Energy Exchange	Transfer energy to a working fluid by direct absorption	Direct-absorption receivers	Heat transfer fluid by direct absorption Sensible heat storage
		Receiver/reactors	Chemical reactions

To reflect these different energy-exchange processes, the following classifications have been made:

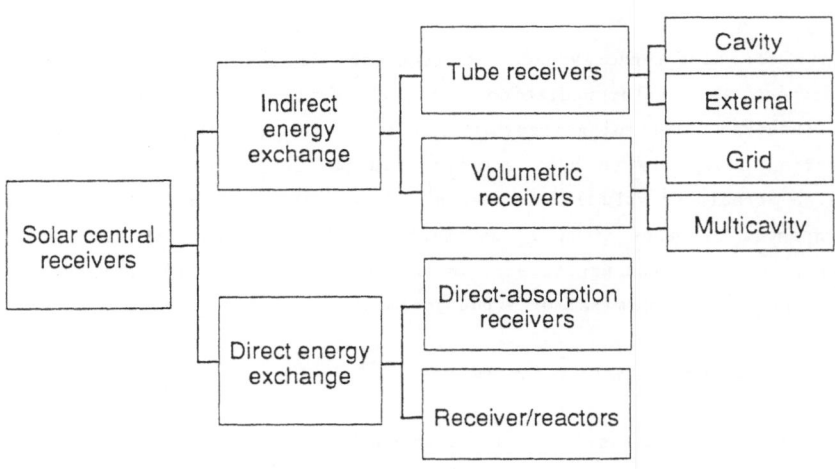

2. Volumetric receivers. The absorber configuration is suited for a three-dimensional geometric arrangement, e.g., a wirepack or multiple-shape design. The basis of this concept is to absorb solar radiation three-dimensionally. Concentrated solar radiation heats the absorber; at the same time, a working fluid flows through it and is heated convectively. When atmospheric air is drawn through the absorber as cooling fluid, convection losses can be avoided almost entirely. A closed loop could be used, in which case the receiver would be equipped with a window.

Grid volumetric receivers contain a series of relatively thin wires, arranged so that concentrated solar radiation is gradually absorbed as it penetrates the wire arrangement.

Volumetric multicavity receivers are formed by an array of cavities or pores. Concentrated solar radiation heats the inside of the cavities; the working fluid flows through the cavities and is heated by convection from the walls.

1.2.2 Direct-Energy-Exchange Receivers

The basis for direct-energy-exchange receivers is the direct absorption of highly concentrated solar radiation, i.e., absorption directly by the working fluid without an intervening absorbing surface. A direct-absorption method of transferring energy is to heat small, solid particles by letting them drop through an irradiated volume. A "seeded" molten salt film could also be used. Heat transport by means of solid particles can be achieved with or without a chemical reaction of the particles. In receivers/reactors, concentrated solar radiation produces a chemical or a photochemical reaction.

1.2.3 Receiver Identification System

In addition to the classifications described, a knowledge of the type of absorber material and the working fluid used is important in identifying the operating range of a solar receiver. Therefore, the following identification system is proposed:

8

Heat transfer, or working, fluid	Absorber material	Receiver type	Receiver arrangement
- Water/steam	- Ceramic	- Tube	- External
- Salt	- Metallic	- Volumetric	- Cavity
- Sodium		- Direct absorption	
- Gas			
- Chemical reactants			

A typical description of a receiver may read, "This is an external, metallic tube type receiver using molten salt as a working fluid."

1.3 Symbols and Nomenclature

In Table 1-2, we have used symbols from the ISO Standards Handbook [1] whenever units were available to meet the needs of this report. We augmented these units with additional ones, as required. Units are also occasionally defined immediately after they appear in the report, for clarity and for the convenience of the reader. The rules of the International Bureau of Weights and Measures apply to magnitudes not included in this table. Because these published standards are subject to review and revision, any modification of SI standards will also apply to Table 1-2. In the table, wherever the name of a parameter is sufficiently clear to define the symbol, an explicit definition was deemed unnecessary and was omitted. Occasionally, the reader is referred elsewhere in this document for definitions or to a thermodynamics text for definitions of the more fundamental thermodynamic concepts.

1.4 Reference

1. ISO Standards Handbook, International Organization for Standardization, Switzerland, 1982.

Table 1-2. Symbols and Nomenclature

Symbol	Name	Definition	Unit of Measurement	International Symbol for Unit
A	area		square meter	m^2
A	availability	Receiver availability [Eq. (2-47)]	dimensionless	--
b	width		meter	m
c	speed of light	Velocity of propagation of electromagnetic waves in vacuum	meter per second	m/s
c	specific heat capacity	Heat capacity divided by mass	joule per kilogram kelvin	J/(kg K)
c_p	specific heat capacity at constant pressure	Heat capacity, at constant pressure, divided by mass	joule per kilogram kelvin	J/(kg K)
D	diameter		meter	m
d	thickness		meter	m
E	energy		joule	J
E	irradiance	At a point of a surface, the radiant energy flux incident on an element of the surface divided by the area of that element	watt per square meter	W/m^2
Gr	Grashof number	$\dfrac{g\beta l^3 \Delta T}{\nu^2}$	dimensionless	--
H	enthalpy	$H = U + pV$	joule	J
h	specific enthalpy	Enthalpy divided by mass	joule per kilogram	J/kg
h	coefficient of heat transfer	Density of heat flow rate divided by temperature difference	watt per square meter kelvin	$W/(m^2\ K)$
h	height		meter	m

Table 1-2. Symbols and Nomenclature (Continued)

Symbol	Name	Definition	Unit of Measurement	International Symbol for Unit
I_θ	radiant intensity	In a given direction from a source, the radiant energy flux leaving the source, or an element of the source, in an element of solid angle containing the given direction divided by that element of solid angle	watt per steradian	W/sr
k	thermal conductivity	Density of heat flow rate divided by temperature gradient	watt per meter kelvin	W/(m K)
L	radiance	At a point of a surface and in a given direction, the radiant intensity of an element of the surface, divided by the area of the orthogonal projection of this element on a plane perpendicular to the given direction	watt per steradian square meter	W/(sr m^2)
l	length	Characteristic linear dimension of body	meter	m
m	mass		kilogram	kg
m	mass flow rate	Rate at which mass crosses a surface	kilogram per second	kg/s
Nu	Nusselt nuber	$\dfrac{hl}{k}$	dimensionless	--
P	power	Rate of energy (heat) transfer	watt	W
Pr	Prandtl number	$\dfrac{\mu c_p}{k}$	dimensionless	--
p	pressure	Force divided by area	pascal	Pa
Q	heat	(Refer to thermodynamics textbook)	joule	J
R	thermal resistance	Temperature difference divided by heat flow rate	kelvin per watt	K/W

Table 1-2. Symbols and Nomenclature (Continued)

Symbol	Name	Definition	Unit of Measurement	International Symbol for Unit
Re	Reynolds number	$\dfrac{vD\rho}{\mu}$	dimensionless	--
r	radius		meter	m
s	length of path		meter	m
S	entropy	(Refer to thermodynamics textbook)	joule	J
s	specific entropy	Entropy divided by mass	joule per kilogram kelvin	J/(kg K)
T	thermodynamic temperature	$\left(\dfrac{\partial U}{\partial S}\right)_V$	kelvin	K
t	Celsius temperature	$t = T - T_o$, where $T_o = 273.15$ K	degree Celsius	°C
t	time		second	s
U	internal energy	(Refer to thermodynamics textbook)	joule	J
u	specific internal energy	Internal energy divided by mass	joule per kilogram	J/kg
v	velocity		meter per second	m/s
V	volume		cubic meter	m^3
v	specific volume	Volume divided by mass	cubic meter per kilogram	m^3/kg
W	radiant energy	Energy emitted, transferred, or received as radiation	joule	J
w	radiant energy density	Radiant energy in an element of volume, divided by that element	joule per cubic meter	J/m^3
α	thermal diffusivity	$\alpha = \dfrac{k}{\rho\,c_p}$	square meter per second	m^2/s
α	absorptivity, absorptance	Ratio of the flux absorbed to that of the incident flux	dimensionless	--

Table 1-2. Symbols and Nomenclature (Continued)

Symbol	Name	Definition	Unit of Measurement	International Symbol for Unit
β	coefficient of thermal expansion	Ratio of unit volume change per unit temperature change	meter cubed per degree Celsius	$m^3/°C$
ε	emissivity, emittance	Ratio of emitted radiation of a thermal radiator to that of a black body at the same temperature	dimensionless	--
η	dynamic viscosity	The coefficient in $\tau_{xz} = \eta \dfrac{dv_x}{dz}$	pascal second	Pa s
η	efficiency		dimensionless	--
λ	wavelength	Distance in the direction of a periodic wave between two successive points at which the phase is the same (at the same time); $\lambda = \dfrac{c}{\nu}$	meter	m
μ	linear attenuation coefficient, linear extinction coefficient	The relative decrease in concentration of the radiant flux of a collimated beam during traversal of a medium, divided by the length traversed	reciprocal meter	m^{-1}
ν	kinematic viscosity	$\nu = \eta/\rho$	meter squared per second	m^2/s
ρ	density	Mass divided by volume	kilogram per cubic meter	kg/m^3
ρ	reflectance, reflectivity	Ratio of the flux reflected to that of the incident radiation	dimensionless	--
σ	Stefan-Boltzmann constant	5.67×10^{-8}	watt per square meter kelvin to the fourth power	$W/(m^2 \ K^4)$
τ	transmissivity, transmittance	Ratio of the flux transmitted to that of the incident flux	dimensionless	--

13

Table 1-2. Symbols and Nomenclature (Concluded)

Symbol	Name	Definition	Unit of Measurement	International Symbol for Unit
τ	shear stress	Shear force per unit area	Pa	N/m^2
Ψ	radiant energy flux rate	At a given point in space, the radiant energy flux incident on a small sphere divided by the cross-sectional area of that sphere	watt per square meter	W/m^2
Ω	solid angle	The solid angle of a cone is defined as the ratio of the area cut out on a spherical surface (with its center at the apex of that cone) to the square of the radius of the sphere	steradian	sr

Chapter 2

Thermodynamic Performance of Solar Central Receivers: Definitions

Thermodynamic Performance of Solar Central Receivers: Definitions

M. Sánchez
CIEMAT-IER, Spain

With contributions by
F. Rosa, Plataforma Solar de Almeria, Spain

Contents

2.1 Thermodynamic Concepts

2.1.1 Instantaneous Values (Power)

• Incoming Radiant Power to the Field

This is the nominal power received as radiation by the heliostat field.

$$P_i = \sum_{j=1}^{N} E_j\, A_j \ , \qquad\qquad (2\text{-}1)$$

where N = total number of heliostats

E_j = irradiance at heliostat

A_j = area of heliostat j.

• Radiant Power Reflected by the Heliostat Field

This is the power reflected as radiation by the heliostat field.

$$P_f = \sum_{j=1}^{N} \rho_j\, E_j\, FG_j\, A_j \ , \qquad\qquad (2\text{-}2)$$

where N = total number of heliostats

ρ_j = heliostat reflectivity

FG_j = geometrical factor of heliostat j; includes shading, blocking, and cosine factors.

• Attenuated Radiant Power

This is the power attenuated as radiation by the atmosphere between the heliostat field and the receiver.

$$P_\mu = \sum_{j=1}^{N} \rho_j\, E_j\, FG_j\, A_j\, (1 - e^{-\mu l j}) \ , \qquad\qquad (2\text{-}3)$$

where μ = linear attenuation coefficient

lj = optical path between heliostat j and the receiver.

18

- Incident Power onto the Receiver Aperture or Absorbing Surface

  This refers to the power incident as radiation on the receiver absorber or aperture. (Note that P_R specifically excludes spillage.)

$$P_R = \int_A \Psi \, dA ,\qquad (2\text{-}4)$$

  where A = the receiver absorbing surface; when referring to a cavity receiver, A is the receiver aperture.

  Ψ = radiant energy flux rate at dA.

- Radiant Power Spillage

  This is the power incident as radiation on the receiver general area, but not incident on the receiver absorbing surface.

$$P_s = P_f - P_\mu - P_R .\qquad (2\text{-}5)$$

- Radiant Power Reflected by the Receiver

  This is the power reflected as radiation by the receiver.

$$P_\rho = \rho_R \, P_R ,\qquad (2\text{-}6)$$

  where ρ_R = receiver reflectivity.

- Radiant Power Absorbed by the Receiver

  This is the power absorbed as radiation by the receiver.

$$P_\alpha = \alpha \, P_R ,\qquad (2\text{-}7)$$

  where α = receiver absorptivity.

19

• Radiant Power Emitted by the Receiver

This is the power emitted as radiation by the receiver.

$$P_\epsilon = \sum_{j=1}^{N} \epsilon \, \sigma \, (T_R{}^4 - T_{sj}{}^4) \, F_{Rj} \, A_R \, , \qquad (2-8)$$

where N = total number of surfaces that exchange radiant
energy with the receiver

ϵ = receiver emissivity

σ = Stefan-Boltzmann constant

T_R = receiver average temperature

T_{sj} = surface j average temperature

F_{Rj} = view factor between receiver and surface j

A_R = receiver surface.

• Thermal Power Lost by Convection

This is the rate of heat flow lost due to convection with ambient air.

$$P_C = \int_{A_R} h_e(T_w - T_a) \, dA \, , \qquad (2-9)$$

where A_R = receiver surface

h_e = effective heat transfer coefficient

T_w = receiver temperature in dA

T_a = ambient air temperature.

• Thermal Power Lost by Conduction

This is the rate of heat flow lost due to conduction.

$$P_{cd} = \sum_{j=1}^{N} - K(\nabla T)_j \, A_j \, , \qquad (2-10)$$

where N = sections where heat is lost by conduction

K = thermal conductivity

∇T = temperature gradient along section

A_j = area of cross section.

• Total Power Lost by the Receiver

This is the total rate of heat flow lost by the receiver.

$$P_{LT} = P_\varepsilon + P_c + P_{cd} + P_\rho, \text{ and also,}$$

$$P_{LT} = P_L + P_\rho ,$$

(2-11)

where P_{LT} = total receiver losses

P_L = receiver thermal losses.

• Net Thermal Power of the Receiver

This is the rate of heat flow into the working fluid.

$$P_N = P_\alpha - P_L = \int_{T_i}^{T_o} \dot{m} \, c_p \, dT ,$$

(2-12)

where T_i = working fluid inlet temperature

T_o = working fluid outlet temperature

$\dot{m}$ = mass flow rate

c_p = specific heat capacity at constant pressure.

2.1.2 Cumulative Values (Energy)

• Incoming Energy to the Field

This refers to the energy received as radiation by the heliostat field.

$$E_i = \int P_i \, dt ,$$

(2-13)

where P_i = radiant power incoming to the field.

21

• Energy Reflected by the Field

This is the energy reflected as radiation by the heliostat field.

$$E_f = \int P_f \, dt \; , \qquad\qquad (2\text{-}14)$$

where P_f = radiant power reflected by the heliostat field.

• Attenuation Losses

This is the radiant energy attenuated by the atmosphere between the heliostat field and the receiver.

$$E_\mu = \int P_\mu \, dt \; , \qquad\qquad (2\text{-}15)$$

where P_μ = attenuated radiant power.

• Incident Energy onto the Receiver

This is the energy incident as radiation on the receiver.

$$E_R = \int P_R \, dt \; , \qquad\qquad (2\text{-}16)$$

where P_R = incident power into the receiver.

• Spillage

This is the energy that is incident on the receiver general area but that does not intercept the absorbing area of the receiver.

$$E_s = E_f - E_\mu - E_R \; . \qquad\qquad (2\text{-}17)$$

• Energy Reflected by the Receiver

This is the energy reflected by the receiver.

$$E_\rho = \int P_\rho \, dt \, , \qquad\qquad (2\text{-}18)$$

where P_ρ = radiant power reflected by the receiver.

• Energy Absorbed by the Receiver

This is the energy absorbed as radiation by the receiver.

$$E_\alpha = \int P_\alpha \, dt \, , \qquad\qquad (2\text{-}19)$$

where P_α = radiant power absorbed by the receiver.

• Energy Emitted by the Receiver

This is the energy emitted as radiation by the receiver.

$$E_\varepsilon = \int P_\varepsilon \, dt \, , \qquad\qquad (2\text{-}20)$$

where P_ε = radiant power emitted by the receiver.

• Convective Losses

This is the heat lost by receiver convection.

$$E_c = \int P_c \, dt \, , \qquad\qquad (2\text{-}21)$$

where P_c = thermal power lost by convection.

• Conductive Losses

This is the heat lost due to conduction.

$$E_{cd} = \int P_{cd} \, dt \, , \qquad\qquad (2\text{-}22)$$

where P_{cd} = thermal power lost by conduction.

● Total Receiver Losses

This is the total energy lost by the receiver.

$$E_{LT} = E_\rho + E_\epsilon + E_c + E_{cd} \ . \tag{2-23}$$

● Net Energy of the Receiver

This is the heat absorbed by the working fluid.

$$E_N = \int P_N \ dt \ , \tag{2-24}$$

where P_N = net thermal power of the receiver.

2.2 Characteristic Times

● Sunrise: t_{sr}

The time of day at which the upper edge (or limb) of the sun appears to cross the horizon in the morning. (Note: some computer programs give sunrise as the time at which the centroid of the sun crosses the horizon.)

● Startup: t_{su}

The time of day at which the receiver starts delivering net energy.

● Warmup: t_{rn}

The time of day at which the receiver reaches its nominal operating conditions.

● Solar noon: t_{sn}

The time of day at which the sun crosses the meridian of the observer. Also called 12 hours solar time.

24

- Cooldown: t_{ln}

  The time of day at which the receiver deviates from its nominal operating
  conditions and begins cooling down.

- Shutdown: t_{sd}

  The time of day at which the receiver stops delivering net energy.

- Sunset: t_{ss}

  The time of day at which the upper edge of the sun appears to cross the
  horizon in the evening.

Figure 2-1 shows the approximate location of each of these times during the
day.

2.3 Efficiencies

2.3.1 Heliostat Field Efficiency

- Instantaneous:

$$\eta_h = \frac{P_f}{P_i} \; .$$ (2-25)

- Daily (from sunrise to sunset):

$$\bar{\eta}_{hd} = \frac{E_{fd}}{E_{id}} = \frac{\int_{t_{sr}}^{t_{ss}} P_f \, dt}{\int_{t_{sr}}^{t_{ss}} P_i \, dt} \; .$$ (2-26)

- During operation:

$$\bar{\eta}_{ho} = \frac{E_{fo}}{E_{io}} = \frac{\int_{t_{su}}^{t_{sd}} P_f \, dt}{\int_{t_{su}}^{t_{sd}} P_i \, dt} \; .$$ (2-27)

25

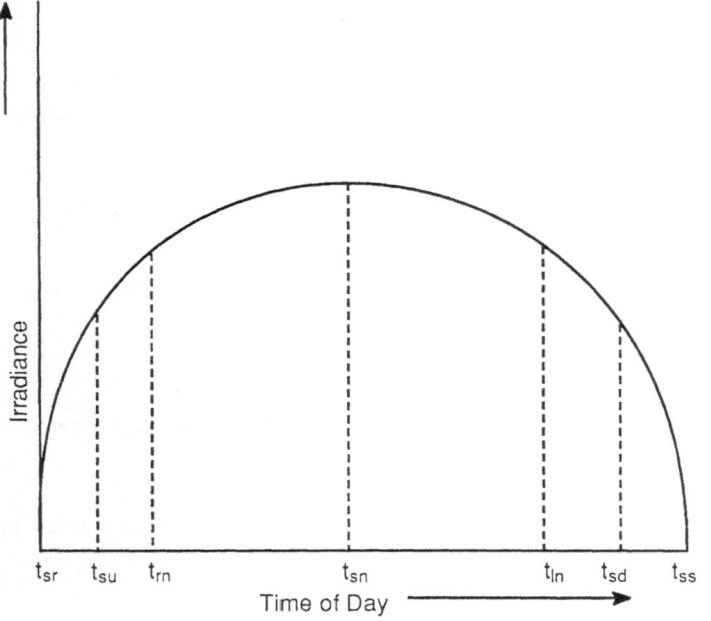

Figure 2-1. Characteristic times

- When the receiver is delivering useful energy:

$$\bar{\eta}_{hu} = \frac{E_{fu}}{E_{iu}} = \frac{\int_{t_{rn}}^{t_{ln}} P_f \, dt}{\int_{t_{rn}}^{t_{ln}} P_i \, dt} \, . \tag{2-28}$$

2.3.2 Efficiency of Power Reaching the Receiver

- Instantaneous:

$$\eta_f = \frac{P_R}{P_f} \, . \tag{2-29}$$

26

• During operation:

$$\bar{\eta}_{fo} = \frac{E_{ro}}{E_{fo}} = \frac{\int_{t_{su}}^{t_{sd}} P_r \, dt}{\int_{t_{su}}^{t_{sd}} P_f \, dt} \ . \tag{2-30}$$

• When the receiver is delivering useful energy:

$$\bar{\eta}_{fu} = \frac{E_{ru}}{E_{fu}} = \frac{\int_{t_{rn}}^{t_{ln}} P_r \, dt}{\int_{t_{rn}}^{t_{ln}} P_f \, dt} \ . \tag{2-31}$$

2.3.3 Interception Efficiency

• Instantaneous:

$$\eta_i = \frac{P_R}{P_f - P_\mu} \ . \tag{2-32}$$

• During operation:

$$\bar{\eta}_{io} = \frac{E_{Ro}}{E_{fo} \, E_{\mu o}} = \frac{\int_{t_{su}}^{t_{sd}} P_R \, dt}{\int_{t_{su}}^{t_{sd}} P_f \, dt - \int_{t_{su}}^{t_{sd}} P_\mu \, dt} \ . \tag{2-33}$$

• When the receiver is delivering useful energy:

$$\bar{\eta}_{iu} = \frac{E_{Ru}}{E_{fu} - E_u} = \frac{\int_{t_{rn}}^{t_{ln}} P_R \, dt}{\int_{t_{rn}}^{t_{ln}} P_f \, dt - \int_{t_{rn}}^{t_{ln}} P_\mu \, dt} \ . \tag{2-34}$$

2.3.4 Spillage

• Instantaneous:

$$\eta_s = \frac{P_s}{P_f - P_\mu} = 1 - \eta_i \; . \tag{2-35}$$

• During operation:

$$\bar{\eta}_{so} = \frac{E_{so}}{E_{fo} - E_{\mu o}} = \frac{\int_{t_{su}}^{t_{sd}} P_s \, dt}{\int_{t_{su}}^{t_{sd}} P_f \, dt - \int_{t_{su}}^{t_{sd}} P_\mu \, dt} = 1 - \bar{\eta}_{io} \; . \tag{2-36}$$

• When the receiver is delivering useful energy:

$$\bar{\eta}_{su} = \frac{E_{su}}{E_{fu} - E_{\mu u}} = \frac{\int_{t_{rn}}^{t_{ln}} P_s \, dt}{\int_{t_{rn}}^{t_{ln}} P_f \, dt - \int_{t_{rn}}^{t_{ln}} P_\mu \, dt} = 1 - \bar{\eta}_{iu} \; . \tag{2-37}$$

2.3.5 Receiver Efficiency

• Instantaneous:

$$\eta_R = \frac{P_N}{P_R} \; . \tag{2-38}$$

• During operation:

$$\bar{\eta}_{Ro} = \frac{E_{No}}{E_{Ro}} = \frac{\int_{t_{su}}^{t_{sd}} P_N \, dt}{\int_{t_{su}}^{t_{sd}} P_R \, dt} \; . \tag{2-39}$$

• When the receiver is delivering useful energy:

$$\bar{\eta}_{Ru} = \frac{E_{Nu}}{E_{Ru}} = \frac{\int_{t_{rn}}^{t_{ln}} P_N \, dt}{\int_{t_{rn}}^{t_{ln}} P_R \, dt} \; . \tag{2-40}$$

28

• Net useful energy:

$$\bar{\eta}_{Rn} = \frac{E_{Nu}}{E_{Ro}} = \frac{\int_{t_{rn}}^{t_{ln}} P_N \, dt}{\int_{t_{su}}^{t_{sd}} P_R \, dt} \cdot \qquad (2\text{-}41)$$

2.3.6 Solar Thermal Conversion Efficiency

Note that P_i is used in the denominators of the equations to denote that these efficiencies are reduced by collection losses due to cosine, shading, and heliostat reflection losses.

• Instantaneous:

$$\eta_g = \frac{P_N}{P_i} = \eta_h \, h_f \, h_R \cdot \qquad (2\text{-}42)$$

• From sunrise to sunset:

$$\bar{\eta}_{gd} = \frac{E_N}{E_{id}} = \frac{\int_{t_{rn}}^{t_{ln}} P_N \, dt}{\int_{t_{sr}}^{t_{ss}} P_i \, dt} \cdot \qquad (2\text{-}43)$$

• During operation:

$$\bar{\eta}_{go} = \frac{E_N}{E_{io}} = \frac{\int_{t_{rn}}^{t_{ln}} P_N \, dt}{\int_{t_{su}}^{t_{sd}} P_i \, dt} \cdot \qquad (2\text{-}44)$$

• When the receiver is delivering useful energy:

$$\bar{\eta}_{gu} = \frac{E_N}{E_{iu}} = \frac{\int_{t_{rn}}^{t_{ln}} P_N \, dt}{\int_{t_{rn}}^{t_{ln}} P_i \, dt} \cdot \qquad (2\text{-}45)$$

29

2.4 Response to Transients

The purpose of all solar central receivers is to deliver power at desirable levels and times, whether the insolation rate varies or not. As a result, solar receivers often require a carefully designed control system to regulate energy output.

For solar thermal receivers, the energy output is in the form of a fluid that is heated to some desired temperature. Typically, controls are designed to keep the outlet temperature constant at changing input power levels. This "quasi-stationary" state is achieved by manipulating the working fluid flow rate. This state must be reached quickly and maintained, while the temperature gradients in the receiver are kept within tolerable limits.

Since variations in irradiance can be large, for example, due to passing clouds, variations in flow rate will be large, too. These variations determine the residence time of the fluid in the receiver, as well as other times that characterize the receiver's response.

Control of the receiver and determination of adequate operation strategies are based on an understanding of its transient behavior. So-called stimulus-response techniques are often used to characterize the transient response of solar receivers. In particular, the "step-input" method has been used for this purpose [1-3].

Based on the configuration and characteristics of the receiver system (i.e., the system composed of receiver, pump, and controls), the transient behavior may be characterized in terms of ratios between variations in the outlet temperature and those in the working fluid flow rate, the heat flux reaching the receiver, and the corresponding time constant.

Since solar receivers are not linear systems, the time constant and the static gain are functions of the operating range; hence, the receiver transient response should be tested under at least three different loads (i.e., low, medium, and high loads).

For convenience in the treatment of the data, a dimensionless temperature defined as reduced outlet temperature, T_r, should be used in mathematical analysis and presentation of results of the transient behavior. T_r is defined [2] as

$$T_{r(t)} = \frac{T_o(t) - T_o(t=0)}{T_{o,ss} - T_o(t=0)} \, , \qquad (2-46)$$

where $T_{o(t)}$ = actual outlet temperature

$T_{o(t=0)}$ = outlet temperature at time zero

$T_{o,ss}$ = outlet temperature when steady state is reached.

Figure 2-2 shows the response function of the reduced outlet temperature for different receivers [2].

The reduced outlet temperature versus time should be plotted in the presentation of results of tests performed, so as to evaluate the time constants. The "forcing function" for the test (e.g., "step-input") should be specified so that it is clear what the receiver is responding to. Also, some consideration should be given to the time required for warmup and some observations should be made about the performance of a receiver system when clouds pass over the heliostat field.

2.5 Availability and Operation

System availability is a good measure of how well a particular system is meeting its intended functions. However, in applications using the sun as an energy source, it is also necessary to quantify actual system operation. These terms are defined below.

2.5.1 Receiver System Availability

Equipment availability is the percentage of a total time period that equipment is available for use, whether it is actually operating or not; it is the ratio of available hours to total hours. Available hours are obtained by subtracting equipment downtime hours from the total-time-period hours.

31

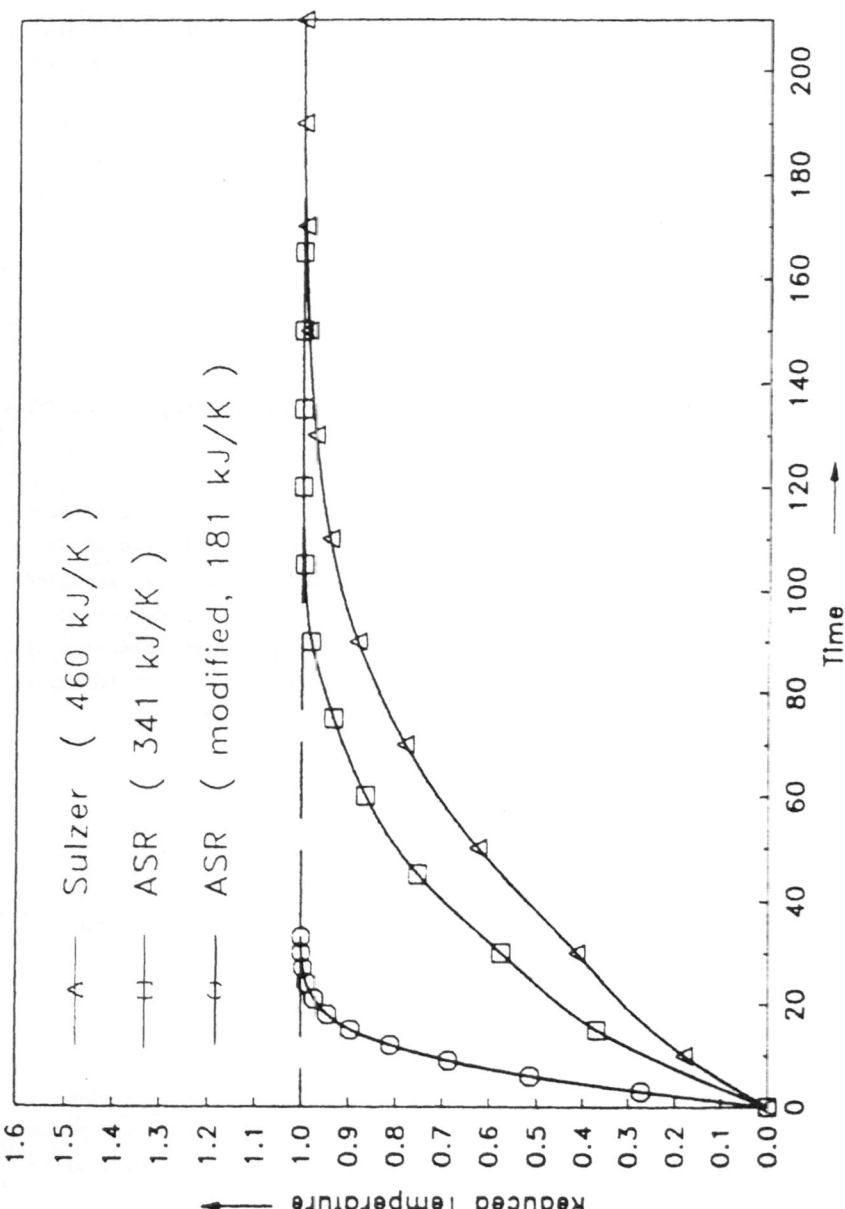

Figure 2-2. Response function of the reduced outlet temperature for three receivers with different thermal inertias [2]

32

Thus, the annual receiver system availability is defined as follows:

$$A_{Ra} = \frac{\text{number of hours available for operation}}{365 \times 24} \qquad (2\text{-}47)$$

$$= 1 - \frac{\text{number of outage hours}}{365 \times 24} \ .$$

2.5.2 Quantification of Receiver Operation

Even with 100% receiver system availability, the receiver cannot be operated
if there is insufficient radiation. Therefore, to quantify receiver operation
it is important to define a receiver threshold radiation level, P_T, as the
minimum level of beam radiation required to ensure proper operation so as to
gain energy under design conditions.

Using P_R, as defined in Eq. (2-4) (the incident power on the receiver), and
P_T, one can estimate the usage factor, UF. This shows the times that the
receiver has been operated in comparison to the times that it could be operated during a given time period. The usage factor is the ratio of the total
number of hours the system has been working at the operating range during a
year to the total number of hours that the level of incident power on the
receiver exceeded the threshold level required for operation. Thus, the
annual usage factor of the receiver is defined as follows:

$$UF = \frac{\text{annual working hours at operating range}}{\text{annual hours where } P_R > P_T} \ . \qquad (2\text{-}48)$$

A second, but equivalent, way to quantify the ability of a central receiver to
accommodate variations in input radiation level by maintaining outlet temperatures sufficient for the conversion system is to use the concept of the turn-
down ratio, or TDR. The TDR is defined as the ratio of the nominal rated
power to the minimum power at which the rated outlet temperature can be
maintained.

2.6 References

1. R. Carmona and J. G. Martin, "The Advanced Sodium Receiver Transient Response," The IEA/SSPS Solar Thermal Power Plants: Facts and Figures, Final Report of the International Test and Evaluation Team, Vol. 1, P. Kesselring and C. S. Selvage, eds., Springer-Verlag, 1986.

2. G. Lemperle, "Non-Steady Thermodynamic Behaviour of Sodium Cooled Receivers. Results cf Calculations and Measurements," Solar Thermal Central Receiver Systems, Proceedings of the Third International Workshop, Vol. 1, M. Becker, ed., Springer-Verlag, 1986.

3. F. Rosa and J. Guerra, "Receiver Transfer Functions of the GAST Metallic Panel," Advances in Solar Energy Technology, Proceedings of the Biennial Congress of the International Solar Energy Society, Hamburg, Vol. 2, W. H. Bloss and F. Pfisterer, eds., Pergamon Press, 1987.

Chapter 3

The Evaluation of Receiver Performance: Recommended Standards

The Evaluation of Receiver Performance: Recommended Standards

M. Carasso
Solar Energy Research Institute, USA

With contributions by
R. Köhne, DLR, FRG
M. Sánchez, CIEMAT-IER, Spain

Contents

3.1 Introduction

In this chapter we define a minimum set of parameters considered necessary in the evaluation of the thermodynamic performance of a solar central receiver that are also useful in economic evaluations. This set of parameters and the procedures recommended for estimating them constitute the standards for evaluating the performance of solar thermal receivers. The set of parameters is defined in this chapter. Recommended standards for estimating them are given in Chapters 4 through 7.

Before we proceed with the definitions of these standards, a short exposition of some of the principles that guided the selection of specific parameters, as well as the preferred procedures for evaluating them, should be useful:

- Whenever there is a choice, measurements are preferred to estimates made by using equations or models.

- The set of parameters for receiver performance should include those indicating "instantaneous" performance (e.g., peak thermal efficiency), as well as the "cumulative" parameters needed to evaluate the ability of a receiver to deliver energy over a period of time (e.g., a year).

- Whenever there is a choice, "direct" measurements of the thermal efficiency are deemed preferable to "indirect" measurements, where "direct" measurements are those involving the measurement of incident power on the receiver aperture, P_R, and that carried in the working fluid, P_N. All other measurements are arbitrarily called "indirect."

- The set of parameters that follows is deemed a minimum set. Users performing receiver evaluations are encouraged to use more than one method to corroborate their estimates.

- In all cases, users are strongly encouraged to perform and report an error analysis, as indicated toward the end of this chapter.

Five parameters constitute the set of standards:

- The Thermal Power Efficiency. This is a measure of the ability of a receiver to convert the incoming radiant flux into energy carried by the working fluid.

- The Daily Thermal Energy Conversion Efficiency. This measures the ability of a receiver to produce energy over a diurnal cycle, including warmup and cooldown. This information provides some insight into the economics of energy production, particularly if the value of this parameter is available for a large number of days.

- Response to Transients. This is a measure of the ability of a receiver to continue to produce power in the presence of rapid fluctuations in the incoming flux, such as those caused by clouds.

- Receiver Availability. This is a measure of the reliability of the receiver, and it depends on the total of planned and forced outage days that occur over a year's operating time.

- The Usage Factor (or its equivalent, the Turn-Down Ratio). This is a measure of the receiver's ability to continue to deliver energy when insolation levels are below the nominal level for which it was designed to operate.

Specific definitions of these standards follow.

3.2 Thermal Efficiency (Short-Term, Steady-State Power)

The thermal efficiency of a receiver is the single most important parameter indicating receiver performance, and the peak value is often quoted and used for quick comparisons among different receivers. Therefore, a plot of the instantaneous thermal efficiency as a function of incident power, $\eta_R = \eta_R(P_R)$, is indispensable (see Figure 3-1). [Equation (2-38) defines η_R as equal to P_N/P_R.]

Note that the receiver themal efficiency can be influenced by the size of a receiver or that of its aperture. Moreover, design changes that can increase the thermal efficiency of a receiver often increase the spillage as well and

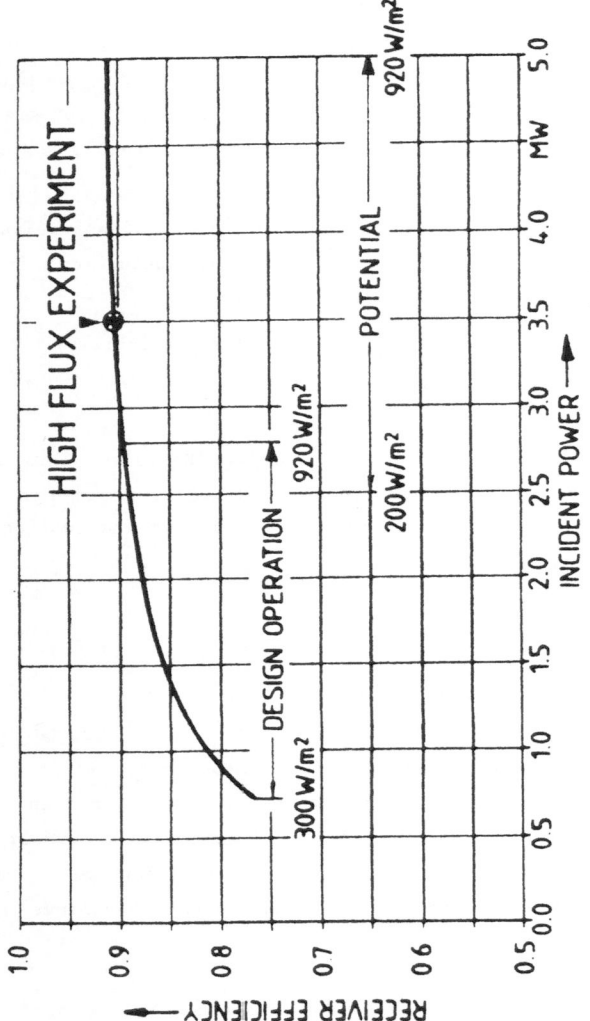

Figure 3-1. Instantaneous receiver efficiency vs. radiant power incident on receiver aperture [1]

result in an overall decrease in system efficiency. Receiver thermal efficiency by itself is therefore not sufficient for comparisons of alternative receiver and system designs.

3.3 Daily Thermal Efficiency (Energy Conversion)

This is a measure of the efficiency of the receiver in converting solar energy, from sunup to sundown, to energy absorbed in the working fluid and available for utilization [as defined in Eq. (2-41)] after losses due to warmup, cooldown, and transients have been accounted for:

$$\bar{\eta}_{Rn} = \frac{E_{Nu}}{E_{Ro}} = \frac{\int_{t_{rn}}^{t_{ln}} P_N \, dt}{\int_{t_{su}}^{t_{sd}} P_R \, dt} \; .$$
(3-1)

Using measurements of daily energy conversion efficiency, typically obtained in different days of the year, one may calculate or otherwise estimate annual efficiencies (see, for example, Figure 3-2).

3.4 Response to Transients

The ability of a receiver to "follow" more or less rapid changes in the incoming flux, such as those due to the passage of clouds--that is, to continue delivering energy at or above a minimum temperature required by the conversion system--is a significant measure of receiver performance. The following equation [as defined in Eq. (2-46)] specifies this parameter:

$$T_r(t) = \frac{T_{o(t)} - T_{o(t=o)}}{T_{o,ss} - T_{o(t=0)}} \; .$$
(3-2)

An example of the abilities of three receivers with different thermal inertias to respond to a step function transient is shown in Figure 2-2.

41

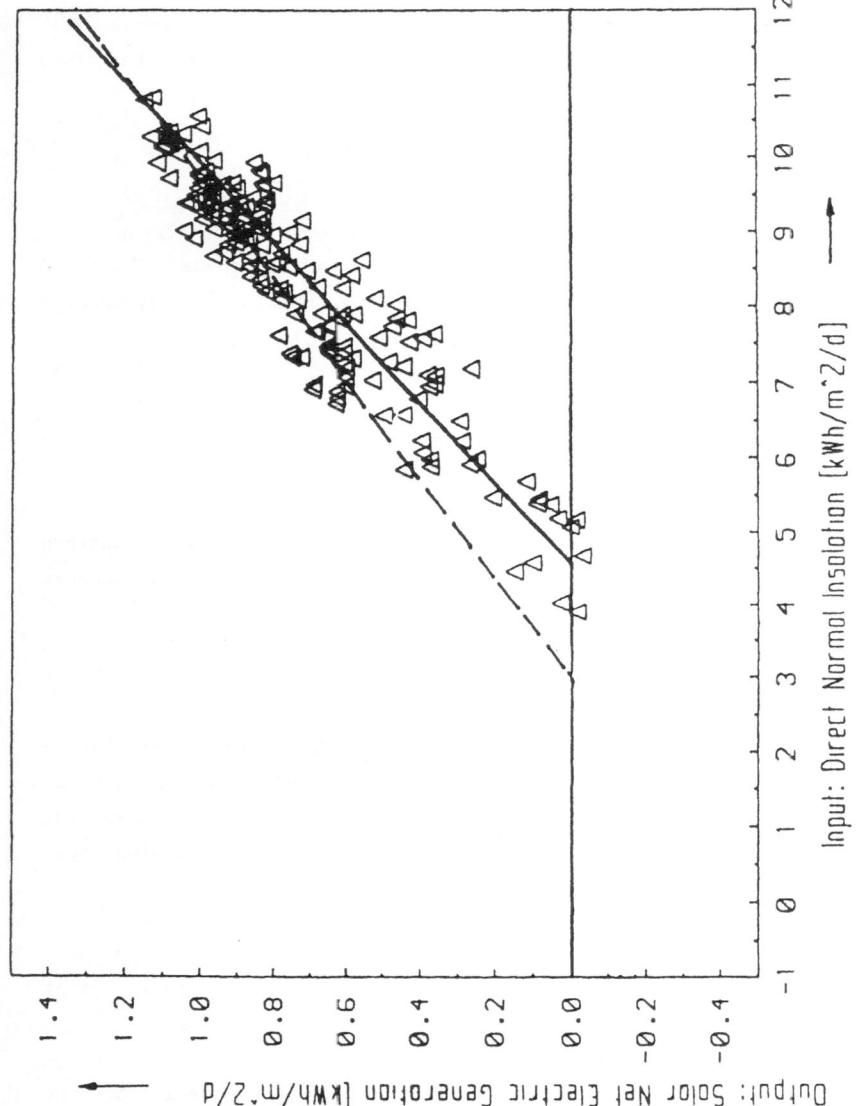

Figure 3-2. Net electric output vs. available solar radiation energy per m² and day [2]

3.5 Receiver Availability

Receiver availability is a measure of how available a receiver is to perform useful work (whether it is utilized or not):

$$A_{Ra} = 1 - \frac{\text{number of outage hours}}{365 \times 24} \, . \qquad (3-3)$$

3.6 The Turn-Down Ratio and the Usage Factor

The turn-down ratio, TDR, is a design parameter for a given receiver. The usage factor, UF, defined in Eq. (2-48), is an equivalent parameter, in that both show the ability of a receiver to continue delivering power when radiation levels are below nominal design values but above a threshold level. The UF requires measurements over a year's time, however, and thus it may be expensive to estimate. It is recommended that, whenever the UF can be measured or estimated, this value be included in reports of receiver performance evaluation.

3.7 Error Estimation Example

An error estimate invariably adds to the quality of reported values as well as to their credibility. It is highly rcommended that an error estimate be made and reported (for example, using "error-bars" in graphs, or numerically) whenever possible.

To know how accurate the measurements of the net thermal power P_N and the radiant power incident on the receiver aperture P_R (and thus the receiver efficiency η_R) are, two errors should be specified:

1. The average relative error
2. The maximum relative error.

The probable average relative error for P_N according to the Gaussian error propagation theorem is

$$\frac{\Delta P_N}{P_N} = \left\{ \left(\frac{\Delta \dot{m}}{\dot{m}}\right)^2 + \left(\frac{\Delta c_p}{c_p}\right)^2 + \left(\frac{\Delta T_o}{T_o - T_i} + \frac{\Delta T_i}{T_o - T_i}\right)^2 \right\}^{1/2} , \qquad (3-4)$$

where $\dot{m}$ = mass flow rate

c_p = specific heat capacity at constant pressure

T_o = heat transfer fluid outlet temperature

T_i = heat transfer fluid inlet temperature.

Thus, the average relative error for the efficiency η_R is

$$\frac{\Delta \eta_R}{\eta_R} = \left\{ \left(\frac{\Delta P_N}{P_N}\right)^2 + \left(\frac{\Delta P_R}{P_R}\right)^2 \right\}^{1/2} , \qquad (3-5)$$

and the maximum relative error is given by

$$\left(\frac{\Delta \eta_R}{\eta_R}\right)_{max} = \frac{\Delta P_N}{P_N} + \frac{\Delta P_R}{P_R} , \qquad (3-6)$$

where P_R = radiant power incident on the receiver.

Since the efficiencies of some receivers (e.g., the Sulzer cavity receiver and the Advanced Sodium Receiver of the 500-kW SSPS Solar Tower Plant of the IEA) differ by less than 5%, it is important to attempt to make measurement errors as small as possible.

Using the relative errors postulated in Chapter 4, one obtains

$$\Delta \dot{m}/\dot{m} = \pm 3\%, \quad \frac{\Delta c_p}{c_p} = \pm 1\%, \quad \frac{\Delta T_{o,i}}{T_o - T_i} \leq \pm 0.5\% \qquad (3-7)$$

$$\frac{\Delta P_R}{P_R} = 4\% , \qquad (3-8)$$

Thus, an average relative error of about ±5% and a maximum relative error of about ±8% should be achievable.

3.8 References

1. W. Schiel, M. Geyer, and R. Carmona, The IEA/SSPS High Flux Experiment, Springer-Verlag, September 1987, p. 178.

2. M. Kiera, W. Meinecke, and P. Wehowsky, Studie zum Vergleich von solaren Turm- und Farmanlagen, Maerz 1990 (private communication).

Chapter 4

Direct Measurements

Direct Measurements

R. Köhne
DLR, FRG

With contributions by
A. Brinner, DLR, FRG
J. Kleih, ZSW, FRG
W. Schiel, Schlaich Bergermann & Partner, FRG

Contents

4.1 Introduction

The receiver efficiency η_R is defined as the ratio of the net thermal power collected by the working fluid of the receiver P_N and the radiant power P_R incident on the receiver aperture. Thus,

$$\eta_R = \frac{P_N}{P_R} .\qquad(4-1)$$

η_R can be evaluated by direct measurement of P_N and P_R.

Measuring P_N may present some minor problems, because of the difficulty of determining the mass flow rate of some working fluids. Measuring P_R is more difficult, and there are many conditions that have to be considered in an accurate and reliable evaluation of efficiency when using only these two direct measurements.

The radiant power P_R, which is reflected by a mirror, heliostat, or heliostat field and incident on a receiver aperture, can be measured directly by either

• radiometers positioned in the aperture plane, or

• special video cameras monitoring a target with a white, diffuse reflecting coating.

The use of radiometers has some disadvantages. Data are obtained only at distinct points, transients cannot be accurately measured, and the loss in accuracy is greater as exposure time increases. The major advantage of this technique, its simplicity, does not outweigh the advantages of a more difficult technique using video camera systems. The latter has a much higher spatial resolution, a fast response time, and higher accuracies. Accuracy especially is of singular importance, however, since the efficiencies of many receivers differ only by a few percentage points.

Measurement techniques and the conditions that must be considered to obtain reliable, accurate results are presented in the sections that follow.

4.2 Measurement of Incident Power

4.2.1 Radiometers

Two different types of radiometers are usually used to map high radiant fluxes:

1. Hycal (or equivalent) radiometers (calorimetric devices)
2. Kendall radiometers.

A Hycal radiometer (usually, series C 1300) consists of a water-cooled copper cavity body that receives radiation on a 16-mm-diameter front face. A circular metal foil with a high-emissivity graphitic coating ($\varepsilon = 0.89$ over a wide spectral range) welded all around the edge of a cavity (diameter ≈ 1.88 mm) in the body is heated relative to the body. The output of this differential copper/constantan thermocouple (foil center and body edges) is directly proportional to the absorbed radiant flux.

Radiometers of this kind are available for measuring maximum fluxes up to about 10 MW/m^2. The response times ($1/e$) for radiometers with upper limits in flux density from 0.15 to 3.5 MW/m^2 range from 250 to 70 ms. Thus, obtaining approximately steady-state values requires a measurement time between 1 and 2 s.

Under laboratory conditions, accuracies of ±3%, a repeatability of ±0.5%, and linearity within ±2% are obtained, resulting in a mean probable error of ±3.5%. The coating of the foil, however, is very sensitive and degrades as a result of outdoor exposure. A misreading of even one radiometer due to overheating in any radiometer arrangement (usually, 10 are installed in a moving bar) can cause a significant reduction in accuracy [1,2].

A Kendall radiometer is a self-calibrating primary absolute cavity radiometer (PACRAD). It consists of a copper receptor with an internally blackened cavity that constitutes a small black body. An electrical calibration heater is located on this black body, which is connected with the main body by a thermopile of 48 Chromel constantan thermocouples. The aperture diameter is 1.2 mm, and the acceptance angle is up to 60° with full cosine response. The radiometer has a wide intensity range (1:500), and flux densities up to 20 MW/m^2 can be measured. The time response is 1.3 s, and a 99.7% response is obtained after 8 s.

The accuracy is at least ±1%, but only when very careful cooling is provided [3]. Accuracy is in fact strongly affected if the flow rate of the cooling water through the radiometer falls below the design value of 4 l/min [4].

Besides the increasing error of the Hycal radiometers after long outdoor exposure, both radiometers have relatively large time constants. Thus, they cannot be used for accurate measurements if there are any rapid transients.

The following systems, consisting of 10 or more radiometers installed on a bar or a crosslike mounting that can be moved across the receiver aperture have been operated successfully:

• Real Time Aperture Flux (RTAF) System, Sandia National Labcratories, N.M., USA

• Heat Flux Distribution (HFD) Bar, DLR, Almeria, Spain [1,2]

• Flux Measuring System (FMS), CIEMAT-IER, Almeria, Spain [5].

4.2.2 VIS-Camera Measuring Systems

The principle of using a video camera (VIS-camera) system to measure the irradiance E at a receiver aperture point (x,y) reflected from a single mirror or heliostat or a complete heliostat field is based on the reflectivity of a Lambertian target screen which is directly brought into the beam. The surface of the target screen has to be Lambertian, i.e., diffuse, and the radiance L (W/m^2 sr) reflected from the surface is independent of the viewing angle and the incident angle of the impinging rays--a prerequisite for this measuring principle. Such a surface can be approximated, for example, by an Al$_2$O$_3$ coating.

A simplified scheme of a VIS-camera system is shown in Figure 4-1. It consists of the VIS-camera, with a resolution of at least 256 × 256 pixels and
256 grey levels (corresponding to an 8-bit A/D converter); a control unit; and
image analysis in which real-time processing is performed (integration, aver-
aging, subtraction, etc.). The video picture is stored in a frame memory that
allows up to 256 images to be summed without any loss in intensity resolution.

Depending on the task (either to measure a single heliostat or a parabolic
mirror at quasi-stationary conditions, or a complete heliostat field without
disturbing the plant's operation), a stationary, mostly water-cooled target
must be used, or a bar moved in a few seconds across the receiver aperture.
In the latter case, the traversing bar is steered by remote control.

Running preferably from right to left, the bar activates several positioning
sensors (preferably one sensor every 10 cm). As each sensor is passed, a
radio signal is transferred to a computer that triggers the camera to take a
picture (40 to 80 ms). The single pictures are composed by a selection code
into one total picture of the flux distribution. These additional features
for determining heliostat-field properties require a fast system and extensive
software. To correlate the results, meteorological data before and after each
measurement are entered into the computer.

This measuring principle is not only restricted to a north field with a cavity
or a "billboard" type of receiver. The radiant power from a circular field
impinging on an external cylindrical receiver can also be measured. This
case, however, is more expensive because the traversing bar has to have a
semicircular shape and has to move vertically. Two devices have to be
installed, and the software for correcting the geometrical image has to be
adapted. The prerequisite that the Lambertian law is still valid [e.g., that
the incidence angles are not too large (see Figure 4-3)], can usually be
assumed.

VIS-camera systems that have been installed to date differ mainly in the type
of applied electro-optical sensor used (the camera tube; e.g., a Vidicon or
CCD); their ability to measure flux distributions of single heliostats or

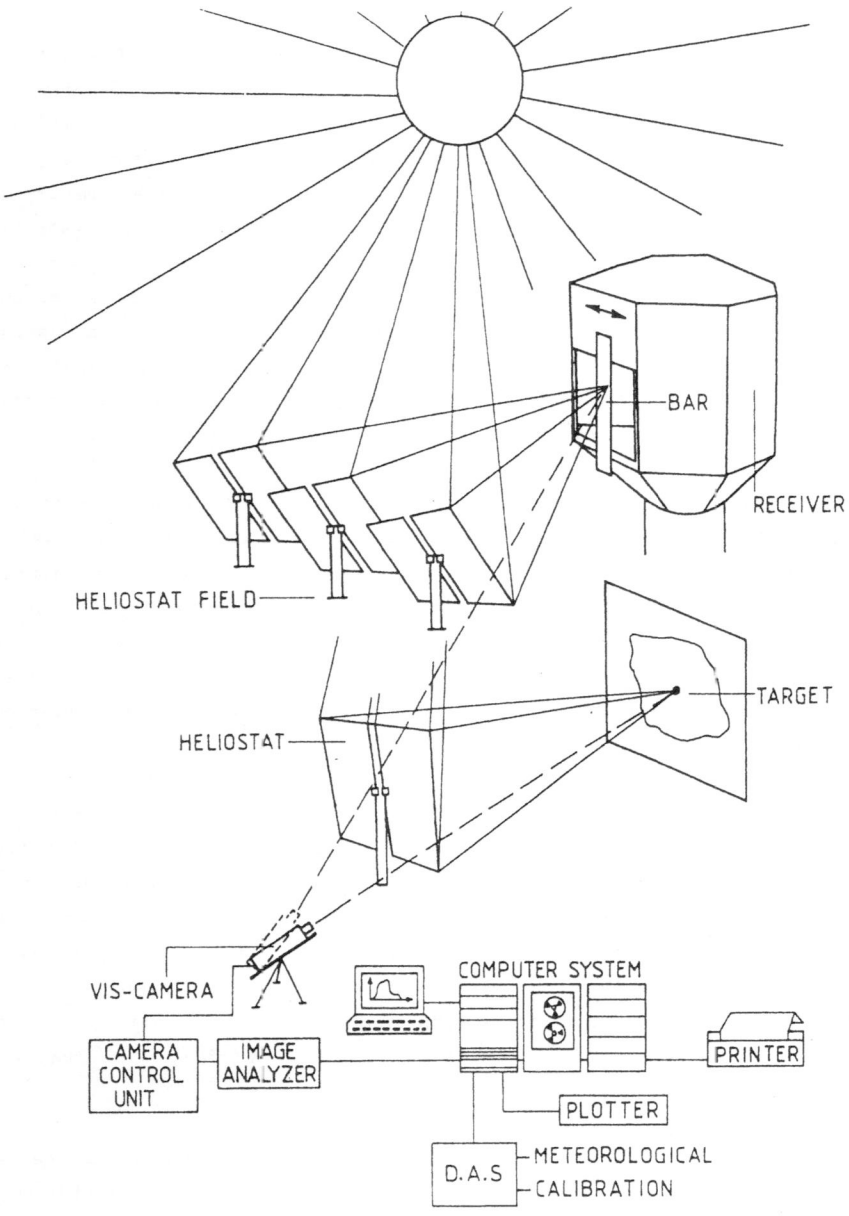

Figure 4-1. Simplified schematic of a VIS-camera system [14]

54

complete fields; time or image resolution, or both; and the accuracy and cost of the software that allows on-line evaluation of various quantities.

The following systems have been operated successfully:

- Beam Characterization System (BCS), Sandia National Laboratories, Albuquerque, NM, USA [6-9]

- High-Intensity Flux Mapper, Solar Energy Research Institute, Golden, CO, USA [10]

- Flux Analyzing System (FAS), the former Eidgenössisches Institut für Reaktortechnik (EIR), now Paul Scherrer Institut (PSI), Winthertur, Switzerland [11,12]

- Heliostat and Receiver Measuring System (HERMES), DFVLR, Stuttgart, West Germany [13-15]

- Two VIS-camera measuring systems developed for the Spanish/German project GAST (Gas Cooled Solar Tower): Measuring System from MBB, Ottobrunn, and Visible Infrared Measuring System (VISIR) from M.A.N., Munich, West Germany [16,17].

Another VIS-camera system for measurements of single heliostats was installed recently at the central receiver test facility at the Weizmann Institute of Science, Rehovot, Israel. A duplication of HERMES will be used for measurements at the Plataforma Solar, Almeria, Spain.

In using these instruments, the following results should be obtained (Figure 4-2):

- Radiant flux density distributions (two- and three-dimensional)

- Isocontour lines

- Total radiant power incident on the receiver aperture.

- When a receiver is used for photosensitive chemical reactions, spectral measurements of each of the quantities above become necessary.

55

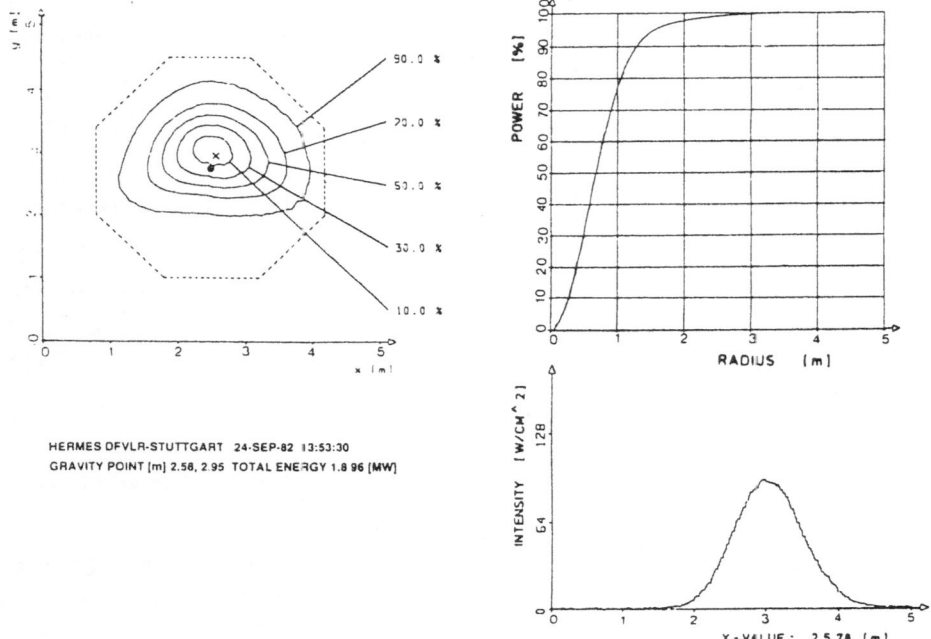

Figure 4-2. Typical plots of the radiant flux incident on a receiver aperture: isocontours, power versus radius, horizontal cross section

From the isocontour lines, the tracking accuracy can be obtained as well as the beam quality, comparing the lines with theoretically computed values (e.g., by HELIOS). Power-versus-radii diagrams are also possible, to obtain mirror, heliostat, or heliostat-field efficiencies.

4.2.2.1 Camera Corrections and Calibrations

The camera signal V (x,y), related to the incident irradiance E (x,y) that is reflected from the point (x,y) of the irradiated target and yields a grey level, is given by

$$V(x,y) = K \, E^{\gamma}(x,y) \, \rho(x,y) \, S(x,y) + N(x,y) \, , \qquad (4-2)$$

where K is the camera constant; $\rho(x,y)$ the reflectivity of the target; $S(x,y)$ the shading factor of the camera due to lenses, filters, etc.; $N(x,y)$ the camera dark current noise, which reduces the dynamic range and is temperature-dependent; and γ is the degree of linearity of the signal output with illumination. It is nearly 1 for a Vidicon tube with a silicon-diode array. A CCD (charge coupled device) sensor also has good linearity and no blooming (focusing problems).

Thus, to obtain the irradiance E in absolute values, the following corrections, measurements, and calibrations must be done:

1. Determination of the dark current.

2. Correction of shading (i.e., the unequal response of the sensor surface).

3. Geometric correction for off-angle viewing.

4. Determination of the linearity of the camera.

5. Determination of the deviation of the target coating from a Lambertian surface and the dependence of reflectivity on wavelength.

6. Measurement of the camera constant K: conversion of measured grey levels into absolute radiant values, i.e., calibration by a primary radiation standard (e.g., a self-calibrating Kendall radiometer).

7. Correction and adaptation with respect to the time response of the calibration standard if fast transients must be measured or a traversing bar is used (e.g., for a heliostat field).

The dark current can be determined easily by placing the lens cap on the camera and recording the output. This should be done before each set of measurements. Since the dark current depends on temperature--almost doubling at each higher 10°C--the camera should be operated in an air-conditioned room.

Shading can be corrected for by using an integrating sphere where the radiation from the exit aperture is nearly perfectly Lambertian. This correction has to be made for each pixel of the active surface of the camera sensor.

Correction of the geometric distortion due to off-angle viewing is done by the cursor interaction on the TV screen, identifying the corners of the receiver aperture and feeding the data into a geometric file for later rectification of the image.

The intensity response of the camera, and thus its linearity, can be determined using an integrating sphere and additional filters. These filters, which are also used for the measurements, must be spectrally neutral because the spectral distribution of the Sun's radiation may change during the day due to differences in atmospheric absorption.

The deviation of the target coating from a Lambertian surface cannot be corrected for. Thus, the only coatings that should be used are those that are nearly Lambertian and whose reflectivities are practically independent of the wavelength. Furthermore, these coatings should be as thin as possible to ensure acceptable heat transfer to the support. Suitable coatings are, for example, 3M NEXTEL Velvet Coating White or Al_2O_3 plasma sprayed coatings. The latter demonstrate, for reflectance, a relatively good wavelength independence--within ±2% and ±3% over the solar spectrum ranging from 0.4 to 2.5 µm. The angular distribution for different incident angles deviates (even for 40°) by only about ±5% compared with that of a Lambertian surface (Figure 4-3) [18].

To determine the camera constant K (i.e., to convert the measured grey levels into absolute values), a 5000- or 10000-sun Kendall radiometer is most suitable because of its high accuracy (±1%) and its large acceptance angle, up to 60° off-axis with full cosine response [4]. The Kendall radiometer must be built into the target, or, in case of a fast traversing bar, into an additional small, water-cooled target (a surface of 0.3 × 0.3 m for a viewing distance of 200 m is recommended) that must have the same coating as the bar

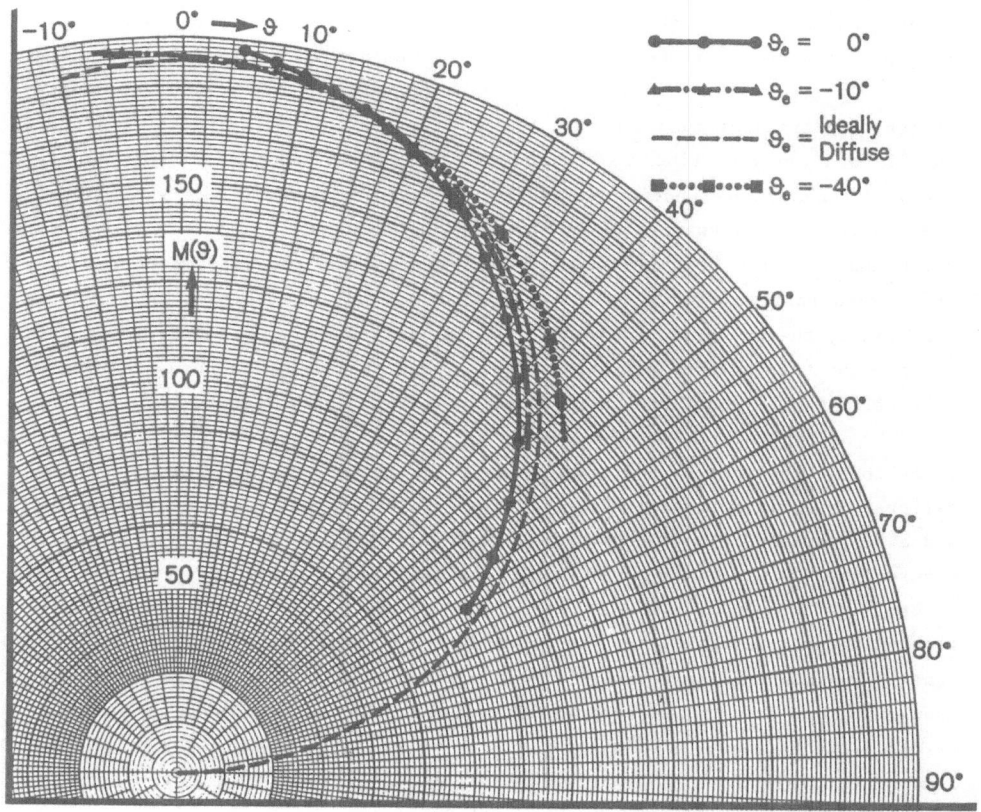

Figure 4-3. Angular distribution of the radiant excitance
M(θ) in arbitrary units of a sample in polar
diagram. Angle of incidence θ_e = 0, 10°, 40° [18].

and that can be moved at different positions across the receiver aperture to
obtain different flux levels. Thus, simultaneous measurements with both the
camera and the radiometer are possible. One must take into consideration that
the flux distribution, even across the small radiometer target, is usually not
uniform. Since the reflectivity at the position of the radiometer (with a
diameter of about 5 cm) drops, the correct camera signal (grey level) must be
interpolated.

Before final absolute calibration, the Kendall radiometer must be adapted to a faster time response if transients are to be measured. The Kendall radiometer has a relatively high time response (99.7% response in 8 s). Thus, an appropriate time-transfer function must be applied. This is described by a differential equation with linear and nonlinear terms that can be derived by applying a defined step function (bright, dark) and monitoring the radiometer's time response. An example of how this works is shown in Figure 4-4 [4], for which the last row of a heliostat field was used and the radiometer remained at a fixed position in the receiver aperture. Curve 1 represents the values measured by the camera, curve 2 the rough data, and curve 3 the corrected data obtained by the radiometer plotted versus the time axis. The response time is well below 0.3 s--short enough to detect relevant flux variations.

To calibrate the camera, one must place the target and the radiometer at three different positions across the receiver aperture: at the edge, at the center, and at a position between them. An example for such a measurement is shown in Figure 4-5, in which the (corrected) Kendall values are plotted versus the camera signals obtained for a combination of two filters used to match different intensity ranges. The deviations are due to the inherent errors of the Kendall radiometer, inaccuracies in the pixel numbers caused by the nonuniform flux distribution (and interpolation) across the target, and superimposed errors from the linearity determination and from the nonideal Lambertian surface of the target. The standard deviation for the lower curve is ±2.5% and for the upper curve ±3.5%, if the small intensity values (Nos. 1 to 8) are disregarded. This is reasonable, because for a low intensity range the lower curve is taken. Thus, an accuracy between ±2% and ±4% can be obtained for the radiant power P_R incident on the receiver aperture.

4.2.2.2 Additional Recommendations

The following two graphic representations should also be included in any paper discussing video camera measuring systems:

- A calibration curve (grey levels plotted versus absolute radiation values of a calibration standard)

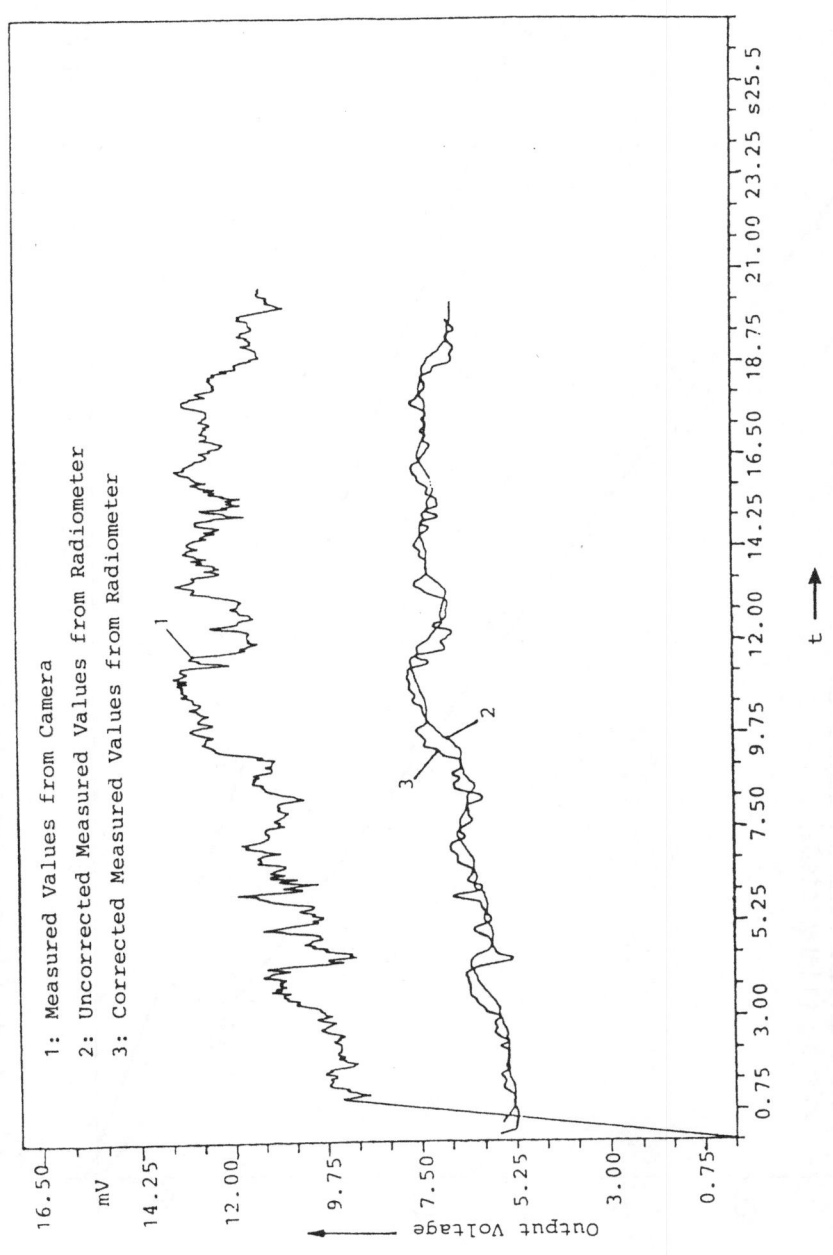

Figure 4-4. Comparison of measured values from VIS-camera and Kendall radiometer [4]

1: Measured Values from Camera
2: Uncorrected Measured Values from Radiometer
3: Corrected Measured Values from Radiometer

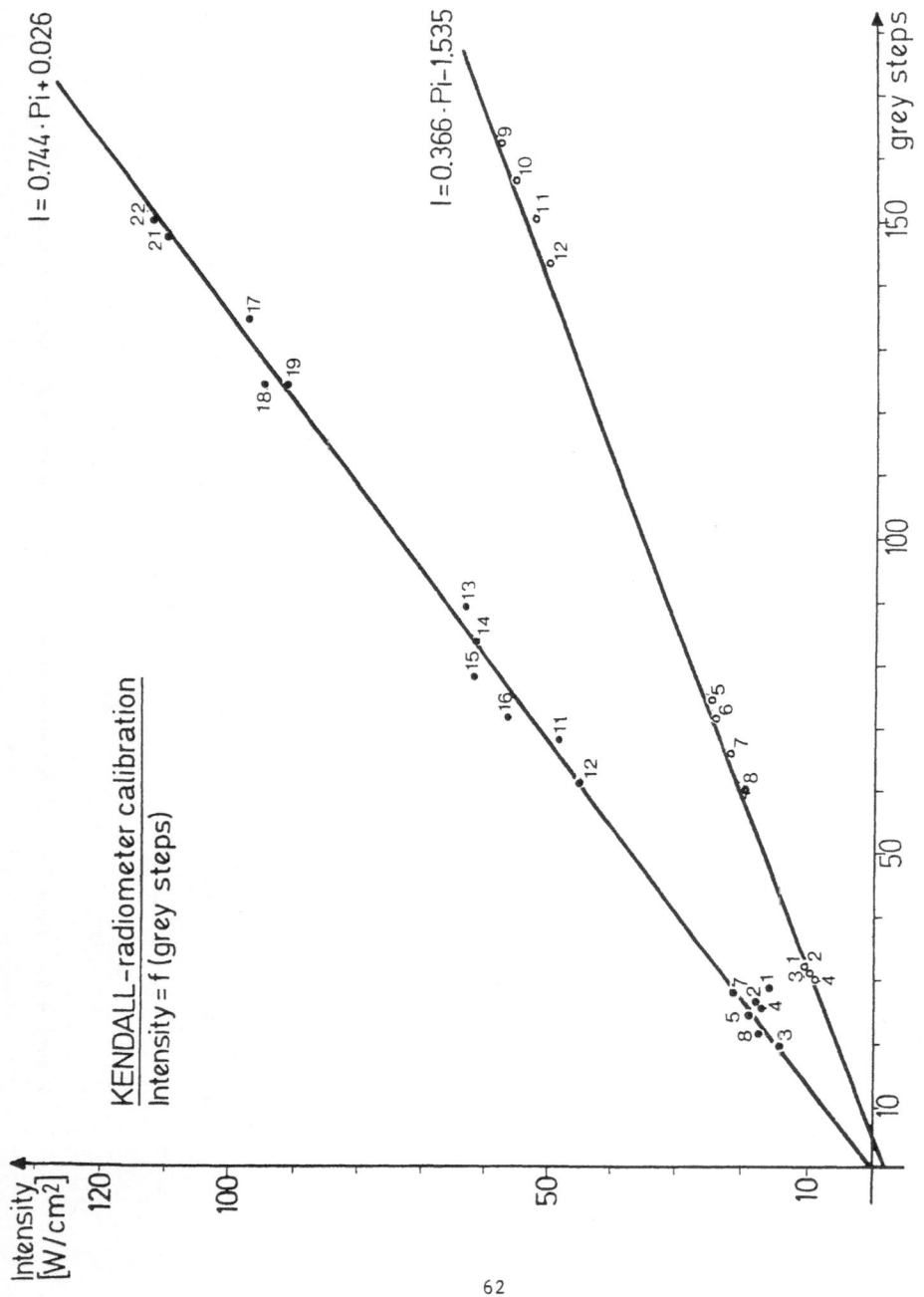

Figure 4-5.   Calibration curve: Absolute intensity values vs. grey levels [4]

62

• A comparison of varying camera levels (qualitative values with the relevant
corrected values of a radiation standard (for transient measurements).

4.3 Measurement of Net Thermal Power

The net thermal power P_N, which is collected by the working fluid in the
receiver, can be determined from the temperature difference between the inlet
and the outlet and the mass flow rate by the following relation:

$$P_N = \dot{m}\, c_p\, \Delta T \; , \qquad\qquad (4-3)$$

where $\dot{m}$ is the mass flow rate, ΔT is the temperature difference $(T_o - T_i)$, and
c_p is the specific heat capacity. Whereas measuring the mass flow rate is
somewhat problematic with respect to accuracy, the temperatures can be
measured easily and accurately if we pay attention to a few conditions.

4.3.1 Temperatures

To measure the temperature of the working fluid, either thermocouples (TC) or
precision resistance thermometers (PRT) are used. The latter have the advantages of stability over time and a higher accuracy due to large changes in
electric resistance at higher temperatures. To prevent adverse effects of
ambient temperature, a three-wire system should be used.

Some difficulties arise concerning temperature measurements in volumetric
receivers, because their outlet diameters often approximate the diameters of
the absorbing volume and thus the aperture diameters. Since the flux distribution across the aperture is usually not uniform, the temperature distribution behind the absorber is very nonuniform. Thus, the mass flow of the
working gas should be adapted to the flux distribution as much as possible,
either by vents of different diameters behind the absorber in wire and foil
receivers or by different pore sizes in ceramic foam receivers.

To obtain reproducible, reliable, accurate values, the following steps should
be followed carefully:

1. Proper Installation

- Temperatures should be measured, respectively, as close as possible to the inlet and outlet of the receiver against the direction of flow, so that the flowing medium hits the sensing part of the sensor first. The temperature sensor should be installed in a location where the flow is well mixed, such as a pipe bend or a specially designed mixing device [Figure 4-6(a)]. The insertion depth should be large enough for the particular PRT: temperature-sensitive length (≈ 25 mm) plus 5 to 10 · d (for fluids and gases, respectively, with d the outer PRT tube diameter).

- A mixing device is an absolute necessity for volumetric receivers. A thermocouple rake should be installed behind to get a fairly reliable average outlet temperature. This may no longer be possible with very large exhaust pipe diameters, however, such as the 5 m diameter for the proposed 30 MW_e PHOEBUS tower plant. Here, indirect measurements at the heat exchanger itself may deliver better results.

- To prevent electromagnetic disturbances, the distance from any power connection should be at least 0.5 m, unless special shielding is provided. Earth circuits must be avoided.

- Between the location of the sensor and the heated walls, sufficient insulation and radiation shielding must be provided to prevent additional heating by more than ±0.5 K [Figure 4-6(b)] [20]. A calculation estimating this correction is usually sufficient.

2. Aging

- Thermocouples, especially, deteriorate as a result of aging. Thus, if it is not explicitly mentioned by the manufacturer, artificial aging by means of mechanical and thermal stresses, such as shaking/heating/cooling cycles, is highly recommended.

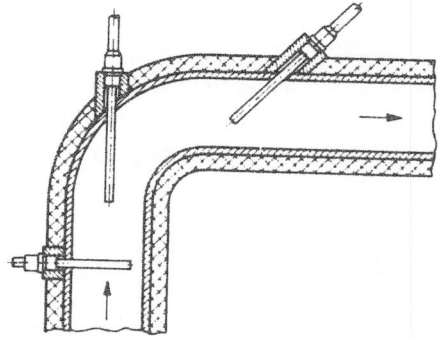

(a)

(b)

Figure 4-6. Installation of thermal sensor in tubes (a); thermocouple
probes for the measurement of gas temperature (shielded)
and for temperature at the tube circumference (b) [20.]

3. Signal Transfer

- The signal transfer through cabling that generally is done over significant distances has to be checked carefully for proper functioning.
There may be large additional errors up to ±1.5 K [19] associated with
signal transfer.

4. Calibration and Recalibration

- Even new thermosensors should be carefully calibrated by a proper calibration arrangement [19]. At a minimum, recalibration is necessary
after each series of measurements.

4.3.2 Heat Capacity and Mass Flow Rate

The heat capacity at constant pressure as a function of temperature must be
known at an accuracy of ±1%. Due to the large temperature spread between the
inlet temperature and the outlet temperature, usually a few hundred kelvins,
the average integral value has to be taken:

$$c_p (T_o, T_i) = \frac{1}{T_o - T_i} \int_{T_i}^{T_o} c_p(T) \, dT \; . \qquad (4-4)$$

To obtain the most reliable and accurate measurement possible of the mass flow
rate, a careful calibration is absolutely necessary before and after a series
of measurements. For liquids, this may be done (if possible) by weighing the
amount of the heat transfer fluid per unit of time (the "litering method")
under operating conditions at identical temperatures and gas cover pressures.

When gas is the working fluid, redundancy in a bypass has to be used instead.
Redundancies are generally recommended for all types of mass-flow-rate measuring devices. The following combinations are possible: for water, turbine
flow meters and the more accurate and less problematic magnetic inductive flow
meters; for gases, turbine flow meters and differential pressure gauges or
meters based on the Coriolis principle (no dependence on medium properties)
and vortex-shedding flow meters (for the latter, the entrance and exit pipes

have to be long enough [≈20 times the diameter] unless a flow straightener is provided); for metallic fluids, two magnetic inductive flow meters (here, exact mounting of the permanent magnets with respect to location on the pipe is very important; a deviation of just 1 mm can cause excessive errors-- greater than 5%) [18].

When two main requirements--careful calibration and a redundant flowmetering-- are fulfilled, an accuracy within ±2% and ±3% can be expected.

Significant difficulty has been encountered with flow measurement in molten-salt systems. In these cases, ΔP-type meters have been employed with pressure transducers that have used silicon-based oil as a remote seal fluid to isolate the transducer from the hot salt. NaK-filled transmitters have proven to be reliable in this application. A vortex-shedding type of flow meter will be tried in a future test.

Although there are no differences between mass flow measurements in small volumetric receivers up to a few hundred kilowatts, and in tube receivers, problems again arise with larger receivers. Especially with open receivers with a significant entrainment of ambient air, a fairly accurate measurement is no longer possible. Here, only an "indirect" measurement of the heat content in the gas by related measurements in the heat exchanger/steam generator itself is possible.

4.4 References

1. M. Becker, J. Bäte, and F. Diessner, Device for the Measurement of Heat Flux Distributions (HFD) near the Receiver Aperture Plane of the Almeria CRS Solar Power Station, Design and Construction Phase, SSPS Technical Report 5/81, 1981.

2. M. Blanco and M. Silva, "Evaluation and Qualification of the HFD Bar," The IEA/SSPS High Flux Experiment, W. Schiel, M. Geyer, and R. Carmona, eds., Springer-Verlag, Berlin, Heidelberg, 1987; pp. 66-79.

3. W. A. Owen, The JPL Flux Mapper, JPL Report 5105-148, 1985; pp. 133-137.

4. A. Brinner, Strahlungsintensitatsmessungen am Solarturmkraftwerk SSPS/CRS der IEA in Spanien, DFVLR FB 85-36, 1985.

5. G. García, General Description of the Flux Measuring System of the Volumetric Receiver, IGA/SSPS International Report, R-15/38GG/1988.

6. E. D. Thalhammer, "Heliostat Beam Characterization System-Update," Advances in Instrumentation (ISA), Vol. 34, No. 2, 1979; pp. 505-521.

7. G. S. Phipps, "Heliostat Beam Characterization System Calibration Technique," ibid.; pp. 523-534.

8. C. L. Mavis, 10 MW$_e$ Solar Thermal Central Receiver Pilot Plant Heliostat and Beam Characterization System Evaluation, Nov. 1981-Dec. 1986, SAND87-8003, Sandia National Laboratories, 1988.

9. J. B. Blackmon, Solar One Beam Characterization System Design Description and Requirements Document, SAND86-8179, Sandia National Laboratories, 1986.

10. T. W. Cannon and H. W. Gaul, A High-Intensity Flux Mapper for Concentrating Solar Collectors, SERI/TP-215-1509, Solar Energy Research Institute, Golden, Colo., 1982.

11. G. von Tobel, Ch. Schelders, and M. Real, Concentrated Solar Flux Measurements at the IEA/SSPS Solar Central Receiver Power Plant, Tabernas, Almeria Spain, SSPS Technical Report 2/82, 1982.

12. W. Durisch and Ch. Schelders, "Radiation Input Measurements with the Flux Analyzing System (F.A.S.)," IEA/SSPS Central Receiver System (CRS) Midterm Workshop, SSPS Technical Report 4/83, 1983; pp. 305-325.

13. W. Schiel, "HERMES Measurements," ibid.; pp. 326-348.

14. W. Schiel, <u>500 kW-Solarturmkraftwerk--Teil I: Mess-Grössen und Mess-System</u>, Brennstoff-Wärme-Kraft (BWK) 6/85, 1985; pp. 266-268.

15. J. Kleih, <u>Beurteilung von konzentrierenden Spiegelsystemen mit Hilfe des Messystems HERMES und des Simulations programms HELIOS</u>, DLR-FB 89-51, 1989.

16. Messerschmid-Bölkow-Blohm, <u>Ausbau und Betrieb der Testeinrichtungen bei MBB</u>, Interatom Report IAS-BT-300000-74, 1983.

17. M.A.N., <u>Messystem VISIR zur Erfassung von Daten im sichtbaren (VIS) und infraroten (IR) Strahlungsbereich</u>, Interatom Report IAS-BT-400000-061, 1984.

18. G. P. Görler, <u>Determination of the Spectral Reflectivity and the Bi-directional Reflectance Characteristics of Some White Surfaces</u>, SSPS Technical Report 6/81, 1981.

19. A. Brinner, <u>IEA/SSPS CRS Calibration Report-Calibration of Relevant Measuring Sensors</u>, SSPS Technical Report 7/84, 1984.

20. Asinel, Interatom, <u>Technologieprogramm GAST Gasgekühltes Sonnenturm-Kraftwerk</u>, Schlussbericht, IAS-BS-010100-182, April 1988.

Chapter 5

Methods of Indirect Evaluation of Receiver Performance

Methods of Indirect Evaluation of Receiver Performance

J. M. Chavez
Sandia National Laboratories, USA

With contributions by
D. Smith, Science Applications International Corporation, USA

Contents

5.1 Introduction

The performance of a central receiver can be determined without direct measurement of incident power on the receiver. Methods of "indirect measurement" make use of measurements and estimates of total thermal losses to evaluate receiver performance indirectly. Methods of determining total losses from a receiver, defined in Eq. (2-11) as P_{LT}, and evaluating receiver performance are described in this chapter. Calculations of individual thermal loss mechanisms are described in the following chapter. A sketch of a receiver with thermal loss mechanisms is shown in Figure 5-1.

"Indirect measurement" methods are evaluation techniques that employ measurements (total thermal losses and absorbed receiver power), calculations, and estimates of parameters that can be utilized in evaluating thermal performance indirectly. Indirect measurement methods are used because direct measurements of many of the parameters (i.e., incident flux, temperatures, and fluid flow rates) may not be easy to obtain, they may be inaccurate, and they may be expensive. Depending on the availability and the specifics of a given receiver, the accuracy of indirect measurement methods may be as good as that of direct measurement methods. Although these methods are not straight-forward, certain phenomena can be exploited to perform the evaluations. Tests with flux on and flux off the receiver can be used in receiver performance evaluation.

5.2 Basic Equations

To understand why and how indirect measurements can be used in evaluating the thermal efficiency of receivers, a brief summary of the basic equations of efficiency evaluation is presented. These basic equations and indirect-measurement methods are discussed in more detail by Boehm, Baker, Kraabel, and Siebers and Kraabel [1-4].

The energy balance on the receiver is given as

$$P_R = P_N + P_{LT} \, , \qquad (5-1)$$

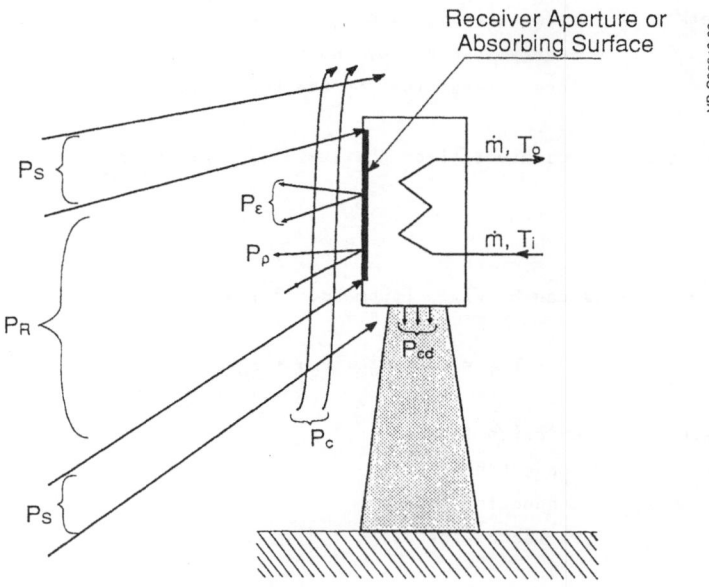

Receiver Aperture or Absorbing Surface

VP-G06248-03

P_R = Incedent power onto the receiver

P_s = Radiant power spillage

P_ρ = Radiant power reflected by the receiver

P_α = Radiant power absorbed by the receiver

P_ε = Radiant power emitted by the receiver

P_c = Thermal power lost by convection

P_{cd} = Thermal power lost by conduction

P_L = Thermal power lost by the receiver

P_N = Net thermal power of the receiver

$P_{LT} = P_L + P_\rho = P_\rho + P_\varepsilon + P_c + P_{cd}$

$P_\alpha = P_L + P_N$

Figure 5-1. Illustration of thermal loss mechanisms from a central receiver

75

where P_R = solar flux incident on the absorber surface

           P_N = energy absorbed by the working fluid

           P_{LT} = total energy lost by the receiver.

The efficiency of the receiver [from Eq. (2-38)] is given as

$$\eta = \frac{P_N}{P_R} \, . \tag{5-2}$$

Total receiver losses can be given [from Eq. (2-11)] as

$$P_{LT} = P_\epsilon + P_c + P_{cd} + P_\rho \, , \tag{5-3}$$

where P_ϵ = emission losses

           P_c = convection losses

           P_{cd} = conduction losses

           P_ρ = reflected losses.

If total receiver losses are measured, or estimated in some way, then the incident flux can be taken out of the efficiency equation by substituting Eq. (5-1) for P_R:

$$\eta_R = \frac{P_N}{P_N + P_\epsilon + P_c + P_{cd} + P_\rho} \tag{5-4}$$

or

$$\eta_R = \frac{1}{1 + P_{LT}/P_N} \, . \tag{5-5}$$

In Eq. (5-5), the measurement of total receiver losses and the energy absorbed by the working fluid become the basis for an estimation of receiver efficiency. Details of evaluating receiver performance by using indirect measurement techniques are given in the following sections.

5.3 Steady-State Receiver Performance Evaluation Using Indirect Measurements

Several methods have been developed and used to experimentally determine thermal losses from a solar receiver using data obtained by operating it in steady state, without direct measurement of the solar power incident on the receiver.

76

For each, the theory underlying the method is described, the mathematical formulation of the method is given, and the literature describing the use of the method is cited.

The operators of a test receiver are generally able to set the receiver heat transfer fluid flow rate, the receiver fluid inlet temperature, and the solar power incident on the receiver. Flow is controlled by a variety of means in the piping system, inlet temperature is set by the system that cools and recycles the fluid, and incident power can be controlled by setting the number of individual heliostats that focus on the receiver. Controlling these independent parameters allows us to control the outlet temperature of the receiver. Maximizing outlet power is usually accomplished by directing all the heliostats to the receiver and then varying the flow to achieve the desired outlet temperature. This ability to control the "operating point" of a test receiver can be exploited, moreover, to determine receiver losses by indirect means.

Receiver losses vary according to the temperature of exposed surfaces and ambient conditions. This temperature (or, more accurately, temperature distribution), in turn, varies with the controlled parameters of receiver operation. In general, the form of this relationship can be specified by an analytical expression based on theoretical considerations. This has the general form

$$P_{LT} = P_R - P_N = f_1(T_s) = f_2(T_i, \dot{m}, P_R) , \qquad (5\text{-}6)$$

where T_s = receiver surface temperature
   T_i = receiver inlet temperature
   $\dot{m}$ = receiver mass flow rate
   f_1, f_2 = general functional relationships.

Several investigators have manipulated these equations mathematically to eliminate the incident solar power (P_R) from the expression. This allows the losses to be determined from experimental data without direct measurement of the incident solar power.

5.3.1 General Methods

In general, two parameters of receiver operation can readily be controlled in an experiment. These are m, the receiver working fluid mass flow rate, which is controlled by adjusting the flow rate, and P_R, the incident solar power, controlled by adjusting the number of heliostats focused on the receiver. As these parameters are manipulated, the receiver temperature and absorbed power change, causing more or less subtle changes in thermal losses. These changes can be observed as changes in absorbed power. The thermal loss can be calculated from these changes.

The working fluid flow rate and receiver incident power can be used to vary the receiver outlet temperature, as they are related by the expression [from Eq. (4-4)]

$$P_N = P_R - P_{LT} = \int_{T_i}^{T_o} \dot{m}\, c_p\, dT \ . \tag{5-7}$$

Methods can be proposed based on procedures that hold one of the parameters of receiver operation (T_o, $\dot{m}$, P_i) constant. In each case, one of the remaining parameters is treated as an independent variable, and the third one is measured as the dependent variable. The most widely used method of this type is the "method of complementary field partitions."

5.3.2 Method of Complementary Heliostat Field Partitions

In the method of complementary heliostat field partitions, a test receiver is operated at constant outlet temperature at three different incident solar power levels. One of these tests is performed at nominal operating conditions, with the full heliostat field focused on the receiver. The other two are performed at approximately half power—one with a uniformly dispersed set of half of the heliostats focused on the receiver, the other with the complementary set of heliostats (those not used in the first set). This ensures that the sum of the incident power of the complementary sets will equal the incident power of the full set (though some minor variations are possible due to changes in solar conditions). The receiver outlet temperature is held

constant by making appropriate adjustments to the receiver flow rate as the
incident power is varied. Ideally, this flow adjustment to maintain a constant outlet temperature can be controlled automatically.

A simplified version of the method was originally proposed by Barron [5]. The
method makes some simplifying assumptions regarding receiver loss. These
assumptions result in errors in the loss determined by the method. An
improved method proposed by Baker [2] was based on the original method, but it
was derived more formally. The assumptions of the original method are identi-
fied, and correction factors are included in the method to account for the
assumptions.

The simplified form of the method makes the assumption that the receiver surface temperature is a function of the coolant temperature only. The thermal
loss P_L can be determined by measuring receiver absorbed thermal power P_N at
three steady-state operating points (subscripts 1, 2, and 3 denote the partial
power, the full power, and the complementary partial power operating points,
respectively):

1. $P_{N,1}$ uses a reduced (usually one-half) incident solar power $P_{R,1}$, by
   focusing half of the heliostats on the receiver and maintaining the
   nominal inlet and outlet coolant temperatures.

2. $P_{N,2}$ employs nominal receiver incident solar power $P_{R,2}$ (the full
   heliostat field) with nominal inlet and outlet coolant temperatures.

3. $P_{N,3}$ uses a reduced incident solar power $P_{R,3}$ by employing the
   complementary set of heliostats relative to the first data point while
   maintaining the same nominal coolant inlet and outlet temperatures.

These data points must be taken in close succession to avoid variations in
solar conditions (insolation) and heliostat field performance. Ideally, these
measurements should be taken as close as possible to solar noon, because
heliostat field performance is most constant at that time. Because the coolant temperature is held constant, the receiver surface temperature is assumed

79

to be constant, and the thermal loss for each operating point P_L is assumed to be the same.

Given these conditions, the incident solar powers are related by

$$P_{R,2} = P_{R,1} + P_{R,3} \cdot \qquad (5\text{-}8)$$

The absorbed power, αP_R [Eq. (2-7)], may be expressed in terms of the net thermal power P_N and the thermal losses P_L as

$$\alpha P_R = P_N + P_L \cdot$$

Substituting the terms for absorbed power and thermal losses [in Eq. (5-8)] yields the expression

$$(P_{N,2} + P_L) = (P_{N,1} + P_L) + (P_{N,3} + P_L) \cdot \qquad (5\text{-}9)$$

This equation may be solved for total thermal losses in terms of the measured absorbed powers:

$$P_L = P_{N,2} - P_{N,1} - P_{N,3} \cdot \qquad (5\text{-}10)$$

This method was employed to determine thermal loss from the Molten Salt Electric Experiment (MSEE) [5] performed at the Central Receiver Test Facility (CRTF) in Albuquerque, N.M. Boehm [6] reported on the results of this test. Three evaluations of thermal loss were made with this method. One of these yielded a physically unrealistic result, but the others provided results in reasonable agreement with theory and with losses estimated by other methods. Boehm noted that the method was quite sensitive to uncertainties in flow measurement and that certain factors (such as time-dependent heliostat field performance and variation of receiver temperature) were not accounted for in the method. Some of these factors are addressed by the improved method.

Baker [2] also noted that the basic (simplified) method of complementary heliostat field partitions was inaccurate for several reasons:

1. The direct-normal solar insolation may be different for each data point.

2. The performance of the heliostat field may be different for each data point as a result of changes in cosine, shading, and blocking losses.

3. The temperature of the front surface of the receiver (and thus the thermal loss) is affected by the incident solar power.

These factors are included in the revised equations. In the improved method, incident solar power is given by the expression in Eq. (2-2):

$$P_f = \sum_{j=1}^{N} P_j \, E_j \, FG_j \, A_j \; . \tag{5-11}$$

Each of these quantities can be measured rather easily, with the exception of the heliostat field performance factor. This factor accounts for the cosine, shading, and blocking losses inherent in the heliostat field, the shadowing losses from the tower, atmospheric attenuation losses, and spillage losses. These quantities vary continuously as a function of time and thus will affect data taken at different points in time. The quantities can be calculated by the use of a heliostat field performance prediction code such as HELIOS [7] or MIRVAL [8].

Two additional ratios must also be provided for this method. These are the ratios of thermal loss at partial operating power to that at full power and are defined as

$$R_{L,1} = P_{L,1}/P_{L,2} \tag{5-12}$$

$$R_{L,3} = P_{L,3}/P_{L,2} \; . \tag{5-13}$$

These ratios contain values that are generally less than 1, since the receiver surface will be hotter and thus the thermal losses will be greater at full-power operation. Estimating values for them, of course, requires that there be some way of estimating the receiver thermal loss, which is the value being measured. These estimates must be calculated by using an independent method to calculate losses. However, the ratios calculated by such methods are expected to be more accurate than the loss calculations themselves. This is because errors in the loss calculations for full and partial power cases are likely to be biased in the same direction. Baker presents methods for estimating thermal loss [2] using analytical techniques. Computer programs that account for much more detail have also been used [9,10].

The net thermal power for the three cases is given by

$$P_{N,1} = \alpha \, P_{f,1} - P_{L,1} \qquad (5\text{-}14)$$

$$P_{N,2} = \alpha \, P_{f,2} - P_{L,2} \qquad (5\text{-}15)$$

$$P_{N,3} = \alpha \, P_{f,3} - P_{L,3} \, , \qquad (5\text{-}16)$$

and input power correction ratios can be defined as

$$R_{I,1} = P_{f,2}/P_{f,1} \qquad (5\text{-}17)$$

$$R_{I,3} = P_{f,2}/P_{f,3} \, . \qquad (5\text{-}18)$$

The thermal loss at nominal power ($P_{L,2}$) may be extracted from these equations as

$$P_{L,2} = \frac{P_{N,2} - (R_{I,1} \times P_{N,1} + R_{I,3} \times P_{N,3})}{(R_{I,1} \times R_{L,1} + R_{I,3} \times R_{L,3}) - 1} \, . \qquad (5\text{-}19)$$

It is noted that if all of the correction ratios ($R_{I,n}$, and $R_{L,n}$) are set equal to one, this equation reduces to Eq. (5-10).

Figure 5-2 illustrates the functional relationship between receiver absorbed power and incident power. Note that the thermal loss will increase with incident power, thus requiring that a correction be made.

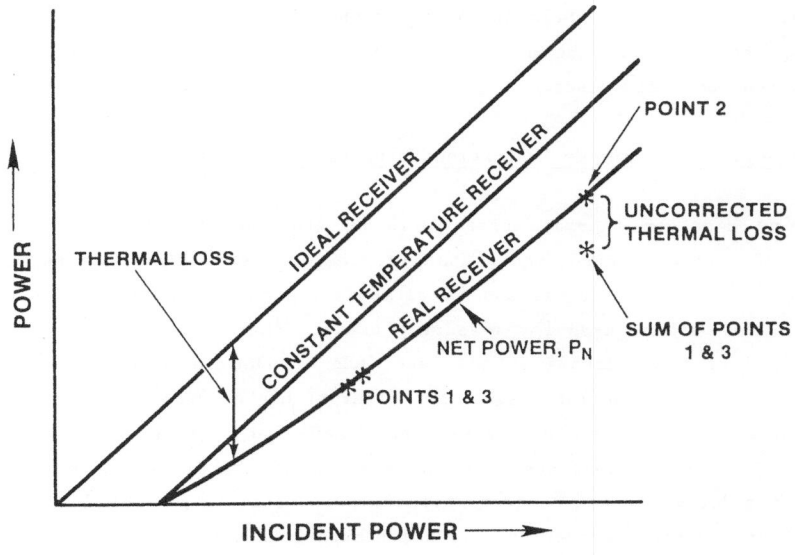

Figure 5-2. Net power vs. incident power

This method has been employed with several test receivers. These include the Solar 1 receiver in Barstow, Calif. [11], the CESA-I Receiver in Almeria, Spain [11], the IEA Small Solar Power Systems (SSPS) Sodium Cavity and External Receivers [2], and the MSS/CTE test receiver at the CRTF in Albuquerque, N.M. [9]. In general, reasonable results were obtained with the method. Mosts of the limitations of the earlier simplified method were addressed. The method retains its sensitivity to the measurement of receiver flow rate, however. Smith [9] reported an error band of ±8% of the receiver power on the losses calculated with this method for a molten-salt receiver. This amounts to an error band of approximately ±100% on the loss measurement. Baker [2] reported errors of approximately ±50% of the measured loss for a sodium receiver. The method also requires considerable analytical support, including estimates of incident power and thermal losses. Despite these weaknesses, the method is the most widely used procedure for indirect evaluation of receiver thermal losses.

5.4 Receiver Performance Evaluation Using Flux-Off Testing

Measuring receiver thermal losses with no flux on the receiver is another method of indirect measurement; that is, the incident solar flux is not used in the experiment. This method is relatively accurate and simple in comparison to the methods used for measuring flux-on thermal losses. However, the flux-off loss measurements do not take into account increases in radiation, convection, and conduction losses that occur at higher operating temperatures. Even though the receiver temperatures are lower than average operating temperatures, Siebers [4] suggests that the results can be scaled. Consequently, flux-off experiments can be quite effective in estimating thermal losses from central receivers. However, the flux-off experiments can be conducted only on receivers in which the inlet fluid temperature is high enough to result in a reasonable temperature difference, such as a molten-salt or sodium receiver. Although these flux-off tests are used to measure total thermal loss, they are often referred to as "convective loss tests." The convective thermal loss can easily be inferred from these tests because radiative losses no longer dominate thermal losses. Because convective losses can be evaluated, a convective heat transfer coefficient can be found and used in scaling thermal losses to higher "operating" temperatures. There are two types of flux-off tests: quasi-steady-state and transient tests.

5.4.1 Steady-State Flux-Off Testing

The idea behind flux-off tests is that if the receiver can be operated without incident solar flux on the absorber, the losses from the receiver can be measured. The total thermal losses are equal to the change in enthalpy of the working fluid.

$$P_L = \dot{m}\, c_p\, \Delta T \ . \qquad (5-20)$$

A plot of typical results is shown in Figure 5-3. In the quasi-steady-state case, the receiver is operated at a constant inlet flow rate and inlet fluid temperature, and the differences between the inlet and outlet temperatures are measured. These tests have also been conducted with the flow directed opposite from normal operation, because often there are limitations on the fluid inlet temperature. If the system can be operated in reverse, the tests can be conducted at higher average receiver temperatures, using the fluid from the hot tank, which more closely approximates operating temperatures.

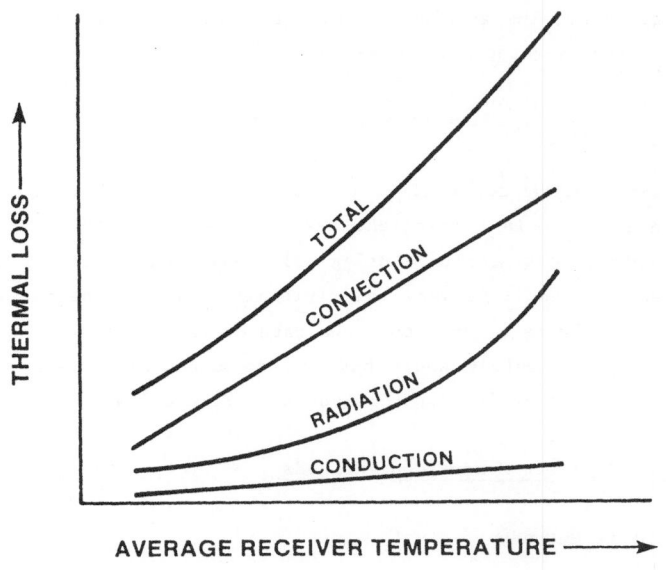

Figure 5-3. Thermal loss from steady-state flux-off testing

Quasi-steady-state, flux-off thermal loss tests have been conducted in all the molten-salt receiver tests [1,5,6,9,12-16] and on the sodium receiver tests conducted at the Plataforma Solar de Almeria [2,3,17]. The mass flow rate and temperature measurements between the fluid inlet and outlet are critical measurements. These measurements affect the uncertainty of the thermal loss; therefore, they must be taken with great care. In addition, because the convective-loss contribution can be the largest fraction, the effects of wind (both speed and direction) on the results of the flux-off thermal loss evaluation are very important and should be taken into consideration. The transport time of the fluid through the receiver may also be important. A correction formula [13] may be used to correct for transport time errors [$\Delta T = T_i(t-\tau) - T_o(t)$, where τ is the transport time].

5.4.2 Transient Flux-Off Testing

The transient response of a receiver can also be used to indirectly measure receiver performance. The gradient in the temperature, at any given time, is proportional to receiver thermal losses:

$$dP_L/dt = \dot{m} \, c_p \, dT/dt \ . \tag{5-21}$$

These tests can provide total thermal losses for a wide range of average receiver temperatures. In a transient test such as this, the receiver would be heated to operating temperature using solar flux; then, the flux would be rapidly removed from the receiver, the fluid remaining in the receiver would be circulated, and the transient cool-down rate of the receiver measured. The thermal mass of the receiver would have to be known for these tests. The basis of the flux-off transient test is shown in Figure 5-4.

5.5 Examples of Indirect Measurement Methods

5.5.1 Steady-State Methods

An example is presented in this section for the steady-state method of complementary field partitions. The method presented here is taken from

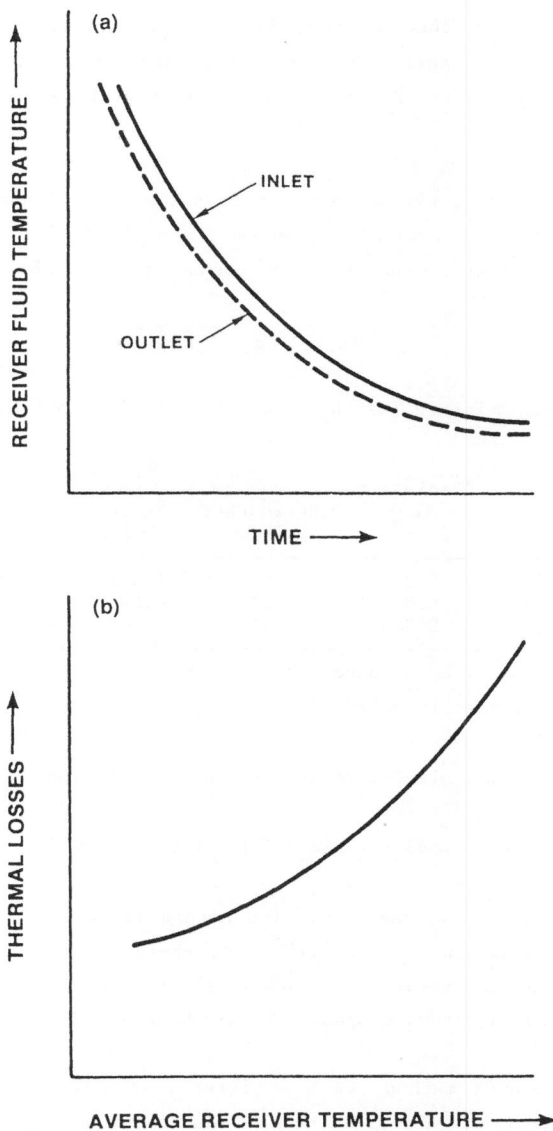

Figure 5-4. Transient flux-off thermal loss test

Baker's evaluation of the sodium-cooled external receiver of the IEA/SSPS project near Almeria, Spain. The data were taken in September 1985 and are documented by Baker [2]. The important experimental data are presented in Table 5-1.

The thermal loss can be evaluated based on the method of complementary field partitions. In the simplified method, the loss is calculated using Eq. (5-10), as modified to indicate the absence of reflection losses,

$$P_L = P_{N,2} - P_{N,1} - P_{N,3} \cdot \qquad (5-22)$$

Table 5-1. Thermal Loss Test Data for IEA/SSPS External Sodium Receiver

Date	Solar Time	Insolation (E) (W/m^2)	Number of Heliostats (N_h)	Absorbed Power Factor (MW$_t$)	Field[a] Performance (FG)	Predicted[b] Thermal Loss (kW$_t$)
1) 9/27/85	10:55	852	59	1.342	0.805	110
2)	11:55	860	110	2.683	0.820	120
3)	12:45	839	51	1.178	0.808	109

[a]Assumed values, not cited in Baker [2].
[b]Using methods presented in Baker [2].

Using the values from Table 5-1 results in the calculation

$$P_L = 2.683 - 1.342 - 1.178 = 0.163 \ (MW_t) \ .$$

The loss calculated using the simplified method is 163 kW. The data in the table, however, show that solar conditions varied significantly during the test. These variations result in errors in the thermal loss evaluated in this manner. The improved method compensates for these errors.

To apply the improved method, we must first calculate the loss ratios and input power correction ratios. These calculations are performed by using Eqs. (5-11), (5-12), (5-13), (5-17), and (5-18):

$$R_{L,1} = P_{L,1}/P_{L,2} = 110/120 = 0.917 \qquad (5-23)$$

$$R_{L,3} = P_{L,3}/P_{L,2} = 109/120 = 0.908 \qquad (5\text{-}24)$$

$$R_{I,1} = P_{f,2}/P_{f,1} = (860)(0.820)/(852)(0.805) = 1.028 \qquad (5\text{-}25)$$

$$R_{I,3} = P_{f,2}/P_{f,1} = (860)(0.820)/(839)(0.808) = 1.040 \; . \qquad (5\text{-}26)$$

These ratios, along with the values in Table 5-1, allow the thermal loss to be calculated using Eq. (5-19):

$$P_{L,2} = \frac{P_{N,2} - (R_{I,1} \times P_{N,1} + R_{I,3} \times P_{N,3})}{(R_{I,1} \times R_{L,1} + R_{I,3} \times R_{L,3})-1} \; . \qquad (5\text{-}19)$$

This expression yields a value of 88 kW$_t$ for the thermal loss in this case. In comparison, the previous, simplified method yielded 163 kW. The reasons for this difference may be inferred from the material presented in section 5.3.1, above.

5.5.2 Flux-Off Methods

5.5.2.1 Flux-Off Quasi-Steady-State Example

Here, we present two examples of quasi-steady-state flux-off testing. One example involves flow in the normal direction and one involves reverse flow.

Normal-Flow Conditions

An example of a steady-state flux-off test with flow in the normal direction is given by tests on the molten-salt cavity receiver of the "Category B" receiver testing conducted at the CRTF in Albuquerque, N.M., in 1987-1988 [9]. In this test, a large number of flux-off tests were conducted, both with the cavity door closed--to measure conduction losses--and with the door open--to measure total thermal losses. An example of the data from the Category B tests is shown in Table 5-2. Note that the data are time-averaged. A large number of data points were taken to evaluate the loss statistically. The effect of wind was also measured.

Table 5-2.   Example of a Flux-Off Thermal Loss
             Test from Category B Receiver
             Tests [9]

T_{in} = 298.8°C T_{out} = 270°C $\dot{m}$ = 10,800 kg/h	
c_{pavg} = 1.532 kW s/kg°C $P_L = \dot{m}c_p\Delta T$ = 132.4 kW	

The uncertainty of these results was calculated to be approximately ±50%, based on mass flow and temperature uncertainties, despite the fact that the mass flow rate throughout the flux-off loss testing was inferred from the change in the pump sump, with flow into the sump shut off. The temperature measurements were a problem until a differential thermocouple was installed. These are a few examples of problems associated with flux-off loss testing. Note that the results presented here are for an average receiver temperature of 280°C, as opposed to the average operating temperature of approximately 425°C. The flux-off tests could not be run at higher temperatures because of temperature limitations on the pump.

The emission losses can easily be calculated for this test, enabling the convective losses to be determined. The results can be extrapolated to operating temperatures to determine receiver performance.

Reverse-Flow Mode

The example for flux-off measurements with flow in the reverse mode is taken from the testing of the Advanced Sodium Receiver (ASR) [17] conducted at the Plataforma Solar de Almeria, Spain. These tests for flux-off in the reverse-flow mode were conducted in addition to the tests with flow in the normal mode. Tests in the normal-flow mode were conducted at fluid inlet temperatures of 200°-300°C, and the reverse-flow tests were conducted at 375°C (normal operating temperature is 550°C).

Since the pump on the ASR did not operate in the reverse mode, the flow rate had to be manually controlled by regulating the difference in argon pressure between the hot and cold tanks. A reverse-flow test should be done at the end of the day, if possible, to give the hot tank the maximum time possible to reach a uniform temperature.

Using the data from both the normal-flow and reverse-flow tests, investigators found the thermal loss from the ASR to fit the polynomial:

$$P_L = 34.6 - 0.18 \, T_m + 0.0015 \, T_m^2 \, , \qquad\qquad (5\text{-}27)$$

where T_m = the average temperature difference between the receiver and the ambient temperature.

These tests provide loss data over a wide range of mean receiver temperatures, from 172°-377°C.

5.5.2.2 Flux-Off Transient Thermal Loss Example

An example of a flux-off transient thermal loss test can be provided by the testing of the Themis molten-salt cavity receiver [14]. This is the only place in the literature where this type of test is documented.

In this test, the receiver was run up to its operating temperature of 450°C. Once steady-state operation was reached, the valves isolating the molten salt in the receiver were closed simultaneously and flux removed from the receiver. The isolated salt in the receiver was continuously cycled through the receiver as it cooled down as a result of thermal losses. Since it took approximately 4 h for the receiver to "cool down" to 250°C, the ambient conditions could change significantly. In addition, the reports indicate that this test was not easy to conduct. Results for the transient flux-off thermal loss test are shown in Figures 5-5 and 5-6. Figure 5-5 shows the change in receiver inlet and outlet temperatures as the receiver cools down (that is, dT/dt). Figure 5-6 shows the results of calculating thermal losses. Convection losses are derived by using thermal loss measurements and radiation losses [14].

91

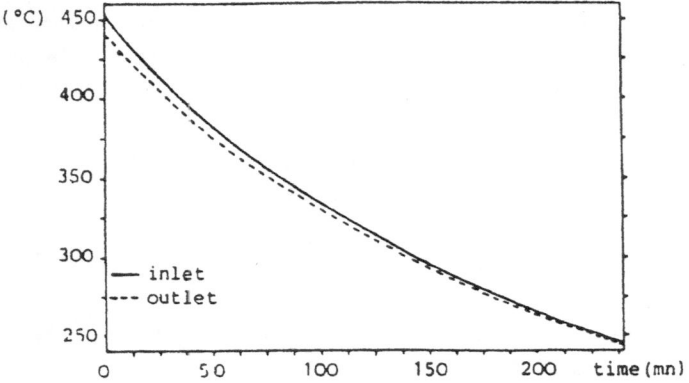

Figure 5-5. Transient flux-off thermal loss test [14]

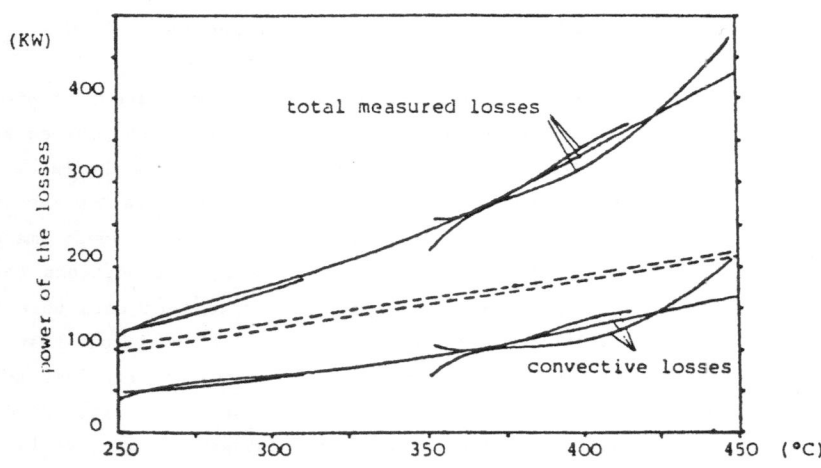

Figure 5-6. Transient flux-off thermal loss test: total thermal losses [14]

5.6 References

1. R. F. Boehm, Review of Thermal Loss Evaluations of Solar Central Receivers, SAND85-8019, Sandia National Laboratories, Albuquerque, NM, April 1986.

2. A. F. Baker, International Energy Agency (IEA) Small Solar Power Systems (SSPS) Sodium Cavity and External Receiver Performance Comparison, SAND87-8021, Sandia National Laboratories, Albuquerque, NM, October 1987.

3. J. Kraabel, Receiver Efficiency Calculation Methods, CRS Fall Equinox Measurement Campaign, 1982, IEA-SSPS DOC R-77/82 JK 3222, September 1982.

4. D. L. Siebers and J. S. Kraabel, Estimating Convective Energy Losses from Solar Central Receivers, SAND84-8717, Sandia National Laboratories, Albuquerque, NM, April 1984.

5. R. J. Holl, D. R. Barron, and S. A. Saloff, Molten Salt Electric Experiment, Research Project RP2302-2, Electric Power Research Institute, Palo Alto, CA, 1987.

6. R. Boehm, H. Nakhaie, and N. Bergan, "A Flux-On Method for Determining Thermal Losses from Solar Central Receivers," American Society of Mechanical Engineers (ASME) paper.

7. C. Vittitoe and F. Biggs, A Users Guide to HELIOS: A Computer Program for Modeling the Optical Behavior of Reflecting Solar Concentrators, SAND82-1562, Sandia National Laboratories, Albuquerque, NM, 1981.

8. A Users Guide for MIRVAL: A Computer Code for Comparing Designs of Heliostats--Receiver Optics for Central Receiver Solar Power Plants, SAND77-8280, Sandia National Laboratories, Albuquerque, NM, 1979.

9. D. C. Smith and J. M. Chavez, A Final Report on the Phase I Testing of a Molten-Salt Cavity Receiver, SAND87-2290, Sandia National Laboratories, Albuquerque, NM, and Livermore, CA, 1988.

10. R. Skocypec and V. Romero, "Thermal Modeling of Solar Central Receiver Cavities," Proceedings of the Tenth Annual ASME Solar Energy Conference, April 10-14, 1988, Denver, CO, L. M. Murphy and T. Mancini, eds., p. 215.

11. A. F. Baker, S. E. Faas, L. G. Radosevich, A. C. Skinrood, J. Peire, M. Castro, and J. L. Presa, U.S.-Spain Evaluation of the Solar One and CESA-I Receiver and Storage Systems, SAND88-8262, Sandia National Laboratories, Albuquerque, NM, and Livermore, CA, 1989.

12. K. J. Beninga and T. Buna, "In Situ Evaluation of Convective Losses in Molten Salt Receivers," Proceedings of the American Solar Energy Society Meeting, 1984.

13. J. W. Grossman, et al., "Phase II Testing of a Molten Salt Central Receiver," 4th International Solar Energy Conference, Santa Fe, N.M., June 1988.

14. B. Bonduelle and A. M. Cazin-Bourguinon, "Themis Receiver: Thermal Losses and Performance," Proceedings of the 3rd International Workshop, Konstanz, Germany, June 23-27, 1986, Solar Thermal Central Receiver Systems, Vol. 1, M. Becker, ed., Cologne, Germany.

15. A. Amri, M. Izygon, and B. Tedjiza, Central Receiver Plant Evaluation, III: Themis Receiver Subsystem Evaluation, SAND88-8101, Sandia National Laboratories contractor report, Albuquerque, NM, February 1988.

16. R. F. Boehm, H. Nakhaie, and D. Berg, Jr., "Heat Loss Experiments on the Category B Solar Receiver," Proceedings of the Tenth Annual ASME Solar Energy Conference, Denver, CO, April 10-14, 1988.

17. R. Carmona, M. Sanchez, and H. Jacobs, "Evaluation of Advanced Sodium Receiver Losses," Proceedings of the 3rd International Workshop, Konstanz, Germany, June 23-27, 1986, Solar Thermal Central Receiver Systems, Vol. 1, M. Becker, ed., Cologne, Germany.

Chapter 6

Evaluation of Individual Thermal Losses

Evaluation of Individual Thermal Losses

D. C. Smith
Science Applications International Corp., USA

With contributions by
J. M. Chavez, Sandia National Laboratories, USA

Contents

6.1 Evaluating Individual Thermal Losses

In this chapter, methods and examples for measurements of four loss mechanisms
(conduction, emission, convection, and reflection) are discussed.

The magnitude of thermal losses varies, and it depends on the receiver type
(e.g., salt-in-tube or volumetric receiver), geometry, and size. Although the
proportion of different thermal losses can vary greatly, depending on the
geometry and the type of the receiver, representative values for the relative
proportions of the different mechanisms in a salt-in-tube cavity receiver are
5% conduction, 50% emission, 30% convection, and 15% reflection losses.

6.2 Conduction Losses

A typical solar central receiver has many surfaces that are hot by virtue of
the receiver's operation, but these play no active role in the absorption of
solar energy.   They include the back of absorber panels, inactive cavity
walls, and process fluid piping. These surfaces are typically insulated to
reduce heat loss to the environment. Although these surfaces can usually be
insulated quite effectively, some heat loss will always occur. Because conduction losses are usually small, they are sometimes neglected, or assumed to
be some small fraction of total losses.   Such a practice should not be
encouraged, because it leaves open the possibility of underestimating them.

The following sections describe ways to determine these losses.   The first
section discusses the calculation of these losses from analytical formulas.
The second discusses experimental ways of determining the magnitude of the
losses.

6.2.1 Description of Conduction Losses

Calculating the conduction loss from insulated surfaces is a relatively
straightforward task. There are several types of losses that must be considered, however. The three types of loss mechanisms that must be considered are
shown in Figure 6-1. These are as follows:

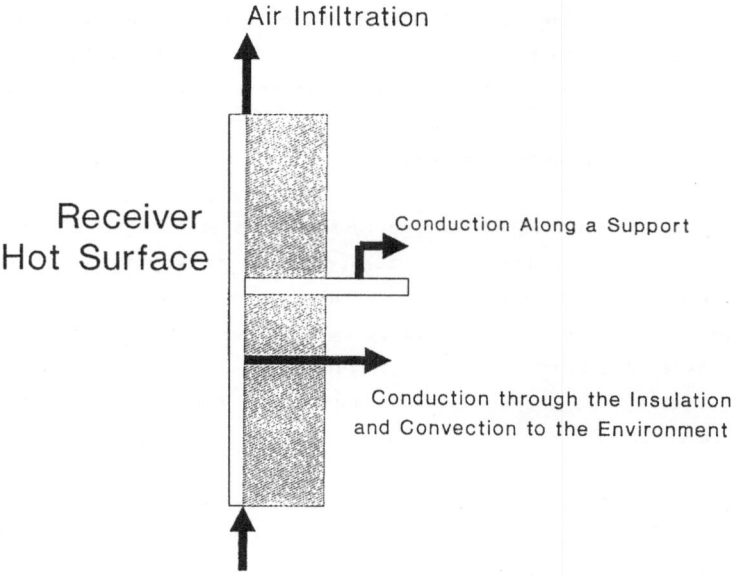

Figure 6-1. Receiver conduction loss mechanisms

- Conduction through a layer of insulating material, and convection from its outer surface to the environment

- Conduction along objects such as structural supports or instrumentation that penetrate the insulation layer

- Infiltration of air through gaps in the insulation to the hot surfaces, and loss of heated air to the environment.

It is necessary for a designer or analyst to consider each of these loss mechanisms carefully. On occasion, one or more of these mechanisms is overlooked or neglected because of the assumption that it is small. In some cases, these loss mechanisms can be significant, so they should always be checked. A simple calculation of these losses may confirm that significant losses are present and that design changes are warranted to deal with them.

Methods that can be used to calculate each of these losses are described in most basic texts on heat transfer. Computer codes are available for calculating heat losses from relatively complex structures and configurations.

6.2.2 Measurement of Conduction Losses

Independent measurement of conduction losses is usually difficult, because the other modes of thermal loss will usually be active any time the conduction mode is active. The other modes, principally convection and emission from the active absorber surface, are typically much larger than the conduction mode. If total losses are measured, it is difficult to divide this value accurately enough to distinguish the conduction component. There are, however, methods that can be employed.

One method for measuring conduction losses that has been employed successfully depends on whether a door is present on a cavity-type receiver. Closing a permanent or temporary door on the receiver aperture can block convection and emission from the active absorber surface. The receiver surface can then be heated by the circulation of the process fluid, or from an independent source. The amount of heat required to maintain the receiver surface at the operating temperature is a good approximation of conduction losses for the cavity enclosure and the door. The door is usually a relatively small portion of a cavity enclosure surface. Door losses can be estimated and deducted from the total. True operating conditions cannot always be determined this way, because such a test is usually isothermal. It can, however, be compared with a calculation of conduction loss for isothermal conditions, allowing verification of the calculation procedure. The calculation procedure can then be applied to the operating conditions with greater confidence.

A similar method might be applied to external receivers and cavity receivers without doors by employing a temporary insulating cover over the active absorber surface, or the cavity aperture.

Specific measurements enabling the calculation of conduction losses may include the following:

- Measurement of external insulation temperatures to see if they match predictions: This can be done using hand-held thermocouples or other devices. Such measurements serve the dual purpose of verifying the calculation and identifying defects in the insulation system by detecting "hot spots" in the insulation envelope. These defects can be taken into account in loss estimation, or remedial action can be taken to eliminate them.

- Measurement of temperatures along structures that penetrate the insulation: Thermocouples can be placed along the length of fins that penetrate the insulation to verify heat-loss estimations for such objects.

- Search for infiltration or exit of air: Often if gaps exist in the insulation, they can be detected as a hot spot on the insulation. As with other defects in the insulation envelope, they can be taken into account in loss estimation, or remedial action can be taken to eliminate them.

6.2.3 Examples of Calculation and Measurement of Conduction Losses

6.2.3.1 Calculation of Conduction Losses

A simple example of the calculation of conduction losses can be taken from a hypothetical example of a flat-plate receiver, as illustrated in Figure 6-2. The receiver surface measures 4 m by 4 m. The back of the receiver is insulated by 10 cm of block insulation. The area behind it is enclosed, so there is no wind, and only natural convection can occur on the surface of the insulation. The insulation is held in place by 3-mm carbon steel pin studs that penetrate through the insulation and extend several centimeters beyond it. The studs are placed in a square pitch, 50 cm apart. Sagging of the insulation between pin studs allows a gap with an average width of 2 mm between the panel and the insulation. The receiver surface operates at a temperature of 500°C.

Calculations:

Conduction. Conduction through the insulation layer can be calculated by dividing the 10-cm layer into three layers, and calculating natural convection heat transfer off its surface (which is typically included as part of the

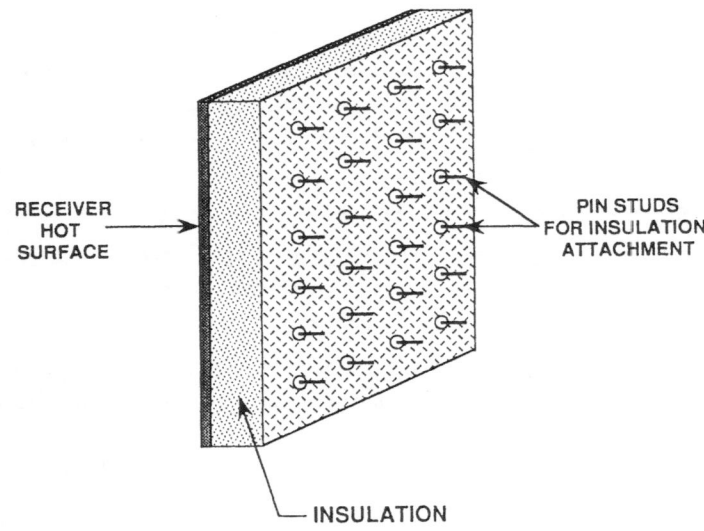

RECEIVER HOT SURFACE

PIN STUDS FOR INSULATION ATTACHMENT

INSULATION

Figure 6-2. A hypothetical flat-plate receiver

"conduction loss mechanisms," for convenience). The conductivity of mineral fiber insulation is illustrated in Figure 6-3. The natural convection heat transfer coefficient for convection to ambient air is given by the expression

$$h_{free} = 1.24(\Delta T)^{1/3} , \qquad (6-1)$$

where

ΔT = the temperature difference between the surface of the insulation and the environment (°C)

h_{free} = the natural convection heat transfer coefficient (W/m^2-°C).

The total heat loss can be calculated using the equation

$$P_{cd} = \Delta T/\Sigma R_{th} , \qquad (6-2)$$

where

P_{cd} = the rate of heat loss

ΔT = the temperature difference between the hot receiver surface and the environment

ΣR_{th} = the total thermal resistance of the insulation system.

The thermal resistance is given by

Insulating layer:

$$R_{th} = t/kA \; ; \qquad\qquad (6-3)$$

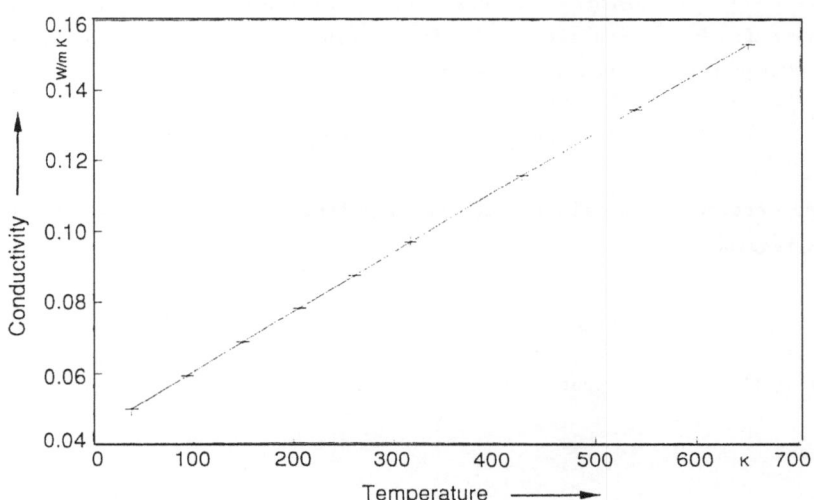

Figure 6-3. Mineral fiber conductivity

103

Surface convection:

$$R_{th} = 1/hA , \qquad\qquad (6\text{-}4)$$

where

    t = the insulation layer thickness
    k = the conductivity of the insulating material
    A = the area of plane surfaces
    h = the heat transfer coefficient for convection.

An iterative calculation is employed to calculate the heat loss, since the conductivity and heat transfer coefficient depend upon temperature, which in turn depends upon heat loss. An initial linear temperature profile is estimated. This is used to calculate conductivity, heat transfer coefficient, and updated temperatures. After several iterations, this procedure converges on a heat loss of 7.07 kW for the $16\text{-}m^2$ insulation surface.

Penetrations. The insulating layer is penetrated by 64 pin studs used to keep the insulation in place. The heat loss from these can be calculated as a composite fin. For the section of fin within the insulation layer, the rate of heat transfer is given by the expression

$$P_{cd} = kA(T_r - T_i)/L_1 . \qquad\qquad (6\text{-}5)$$

For the section of the fin outside the insulation, the heat loss is given by the expression

$$P_{cd} = \sqrt{PhkA}\ (T_i - T_a)\ \tanh(mL_2) \qquad\qquad (6\text{-}6)$$

(assuming that the fin protrudes a long distance through the insulation),

    where

        P_{cd} = the heat loss due to conduction (includes natural convection)
          k = the conductivity of the fin material

104

A = the cross-sectional area of the fin

T_r = the temperature of the receiver where the fin is attached

T_i = the temperature of the fin where it exits the insulation

T_a = the ambient environmental temperature

L_1 = the length of fin inside the insulation

L_2 = the length of fin outside the insulation

P = the fin cross-section perimeter

h = the convection heat transfer coefficient for the fin

$m = \sqrt{hP/kA}$.

Equations (6-5) and (6-6) can be solved simultaneously to obtain the heat loss from the fin. Based on these equations, each pin stud loses 55.9 W, and the total for all 64 studs is 3.5 kW, or almost 50% of the loss through the insulation itself. Clearly this value should not be neglected.

Infiltration. The panel can be expected to grow in length by approximately 2 cm as a result of thermal expansion when it is heated from ambient temperature to 550°C. This may allow gaps in the insulation layer where air can enter and leave the space between the insulation and the receiver panel. The velocity of the air in this gap can be calculated by equating the friction loss to the buoyancy of the hot air in the gap, using the expression

$$v = \sqrt{(2b/fh)(1 - T_a/T_{hot})gh} \ . \qquad (6-7)$$

The heat lost because of this is calculated from the following:

$$P_{inf} = \rho_a v A c_p (T_a - T_{hot}) \ , \qquad (6-8)$$

where

P_{inf} = the infiltration heat loss
ρ_a = the density of air at ambient temperature
v = the velocity of the infiltration air

105

T_{hot} = the temperature of the air at the receiver temperature

h = the height of the receiver

A = the cross-sectional area of the gap behind the panel

b = the width of the gap behind the panel.

Based on this equation, the air velocity in the gap is calculated to be 0.58 m/s, and the infiltration loss is calculated to be 2.9 kW. Again, this value should not be neglected.

6.2.3.2 Measurement of Conduction Losses

The advanced molten-salt cavity receiver [2] tested at the CRTF in Albuquerque, N.M., had a movable door that could be closed in the cavity aperture. The receiver also employed an electric heater within the cavity that was used to heat the receiver panel to the operating temperature with the door closed. (This allowed molten-salt flow to be initiated before sunrise for maximum collection of solar energy.) These heaters consumed 12 kW of electric power. This value is a good measure of conduction loss.

6.3 Emission Losses

Emission losses from central receivers are the thermal energy lost primarily by infrared emission due to the high temperature of the receiver. In evaluating emission losses, the receiver, ground, and sky need to be considered. Given the temperatures of those three surfaces and the emissivity of the receiver, the emission losses can be determined. Because of the scalloped surfaces of tubular receivers, the apparent emissivity is higher than that for a flat plate. Depending on the receiver design, emission losses can account for 30%-70% of the total thermal losses.

6.3.1 Calculation of Emission Losses

It appears that the best methods of estimating emission losses from central receivers are to use computer models (described in Chapter 7.) Anything from simple calculations of emission losses based on average temperatures to elaborate computer models can be used. The standard equation for determining emission heat loss is of the form

$$P_\varepsilon = \sum_{j=1}^{N} \varepsilon\sigma \, (T_R^4 - T_{sj}^4) \, F_{Rj} \, A_R \ , \tag{6-9}$$

where

P_ε = emission heat loss

ε = emissivity of receiver surface

A_R = receiver surface area

T_R = average temperature of emitting surface

T_{sj} = average temperature of surface j

F_{Rj} = view factor between the emitting surface and surface j

$\sum$ = sum over all surfaces.

In the simpler calculation techniques, an average emissivity and receiver temperature are used (for a cavity receiver, this is done at the receiver aperture). In most calculations, the emission exchange between two and five surfaces is generally used, depending on the number of receiver panels. In more complex models, the receiver is divided into zones and simultaneous equations are solved to determine emission losses. The view factor between various zones must be evaluated in the more complex evaluations. The error bounds of these models can be defined and presented with the results. Calculating the emission losses from a billboard or external receiver is much simpler than doing so for a cavity receiver. Since emission losses are closely coupled to convection losses, the more complete computer codes generally include a coupling of emission and convection losses based on calculated surface temperatures (see Chapter 7).

6.3.2 Measurement of Emission Losses

Actual measurements of emission losses are very difficult to obtain. Furthermore, they can greatly increase the cost of a receiver test. Actual measurements have been attempted using infrared camera systems to measure the

107

emission loss of a small area of the receiver. From the emission losses, the surface temperature can be calculated. The accuracy of the emission loss measurement depends on the accuracy of the infrared camera system and the complexity of the receiver. Emission measurements from external receivers are easier to obtain than those from cavity receivers.

6.3.3 Examples of Emission Losses

Calculations:

An example of a relatively simple calculational technique for evaluating emission losses is taken from Jacobs [3] for the evaluation of the Advanced Sodium Receiver (ASR) and the Sulzer sodium receiver. A simplification of Eq. (6-9) is used:

$$P_\epsilon = \sigma \epsilon A (T_R^4 - T_a^4) \ . \tag{6-10}$$

The mean sodium temperature was assumed to be representative of the mean receiver temperature. In this evaluation, the emission losses were estimated by multiplying the power radiated from a black body at the mean sodium temperature to another black body at the ambient temperature multiplied by the remaining products. The emittance of the Pyromark® coated surface was assumed to be 0.9, and the shape factor was taken to be 1.

For specific data, the emission losses were evaluated and plotted. Figure 6-4 shows the emission losses as a function of mean receiver temperature.

Another example that uses a slightly more complex evaluation technique for emission losses is taken from Carmona [4]. In this example, the sky and ground are taken into consideration in the evaluation of emission losses. The equation used is as follows:

$$P_\epsilon = \epsilon \sigma A [0.5 (T_R^4 - T_{sky}^4) + 0.5 (T_R^4 - T_a^4)] \ . \tag{6-11}$$

These results were compared with measurements taken using the HERMES system and proved to agree within a few percentage points.

An example of a more complex method of evaluating emission losses is provided
by the testing of the MSS/CTE molten salt receiver at the CRTF in Albuquerque,
N.M. [2]. In this test the CAVITY computer code [5] was used in calculating
the emission and convection losses from the cavity.

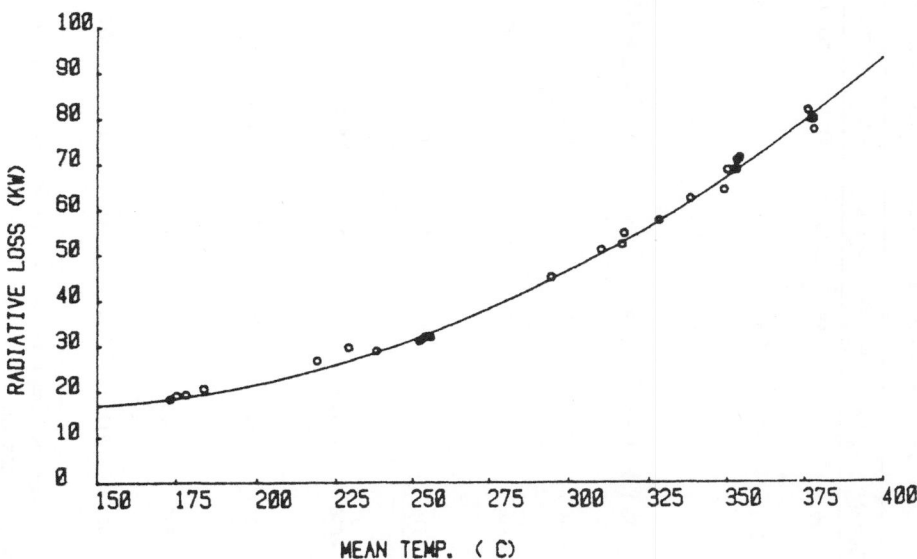

Figure 6-4. Emission losses as a function of receiver temperature [4]

The code does only a single band analysis (i.e., all surfaces are assumed
"grey"). Skocypec and Romero [6] evaluated the effect of convection, nonabsorbing surface absorptivity, and spillage on emission losses from the
MSS/CTE receiver. The results are shown in Table 6-1.

For a cavity receiver, this code is an excellent method of evaluating emission
losses.

109

Table 6-1. Results of the Cavity Model Analysis [6]

Case	P_R(kW)	P_a(kW)	P_c(kW)	P_ϵ(kW)	η(%)	$\bar{T}$(K)	$\dot{m}_c(\frac{kg}{s})$	$\dot{m}_w(\frac{kg}{s})$
Noon, 100% Field Global Convection	4836	4426	149	261	91.5	760.4	5.718	5.723
Noon, 100% Field 1/2 × Global Convection	4836	4479	84	273	92.6	801.8	5.786	5.791
Noon, 100% Field 2 × Global Convection	4836	4331	268	237	89.5	727.0	5.595	5.600
Noon, 100% Field No Convection	4836	4544	0	292	94.0[a]	849.8	5.870	5.876
Noon, 100% Field Local Convection	4836	4416	160	260	91.3	757.5	5.704	5.709
Noon, 100% Field Global Conv., $\epsilon_{kaowool}$ = 0.1	4836	4425	153	258	91.5	767.8	5.718	5.722
Noon, 100% Field Global Conv., $\epsilon_{kaowool}$ = 0.7	4836	4415	165	256	91.3	795.2	5.704	5.709
Noon, 100% Field Global Conv., No Internal Spillage	4836	4309	124	220	89.1[b]	701.4	5.572	5.572
Noon, 100% Field Global Conv., Pyromark Conductivity Included	4836	4414	153	269	91.3	768.6	5.703	5.707
11:00 a.m., 100% Field Global Convection	4702	4303	146	253	91.05	753.4	5.587	5.535
Noon, 50% Field Global Convection	2421	2119	122	180	87.5	699.4	2.737	2.740

[a]If global convection losses are included based on $\bar{T}$ = 851 K (192 kW), η = 89.9%.

[b]Predicted internal spillage is 183 kW.

Measurement

In evaluating the ASR tested at the Plataforma Solar, in Almeria, Spain [7],
the HERMES system was used for measuring the emission loss and front surface
temperatures. HERMES is an integrated system (described in Chapter 4); the
measurements taken by the IR camera are stored in software, and the temperature and emission evaluated of each pixel of data. The emission from each
pixel is calculated and is summed to give the total emission loss. An example
of the emission loss data from measurements using the HERMES system is shown
in Figure 6-5.

6.4 Convection Losses

Estimation of convection losses from a solar receiver presents several problems. First, solar receivers are very large and very hot. As a result, natural convection is driven by temperature gradients higher than those explored
experimentally before the early receivers were designed. Therefore, the
standard heat transfer literature did not provide correlations for the range
of conditions in which solar receivers operate. Measurement of convection
loss is difficult, because experimental methods are typically limited to the
determination of total thermal loss, and it is difficult to separate out convection alone.

Some progress has been made in the development of more applicable heat transfer correlations, and novel approaches to the measurement of convection losses
have been advanced. This section summarizes these advances and illustrates
their application to central receiver systems.

6.4.1 Calculation of Convection Losses

The recommended procedure for calculating thermal convection loss from solar
central receivers follows the recommendation of Siebers and Kraabel [8]. This
procedure has gained the most acceptance among receiver designers and
analysts. Other work has been performed in this area [9-13], however, and the
reader is encouraged to investigate these and other sources. The topic of
central receiver convection thermal loss is still developing.

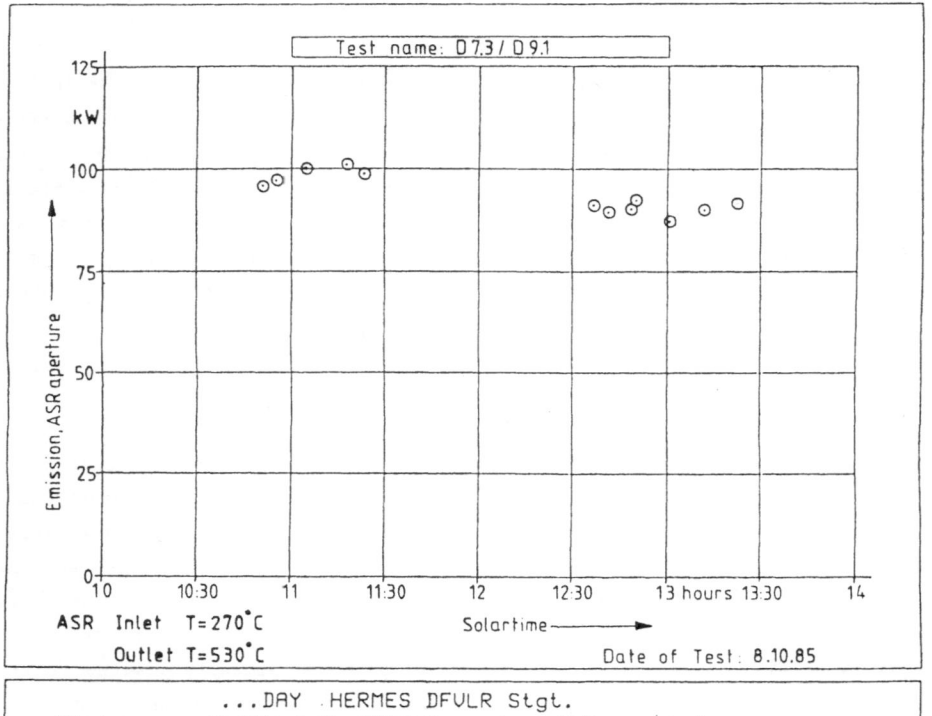

Figure 6-5. Example of emission loss measurements using HERMES [7]

The convection heat loss from a central receiver is given by the general
expression

$$P_c = hA(T_w - T_a) \ , \qquad\qquad (6-12)$$

where

  T_w = the mean receiver wall temperature.

The value of the receiver mean heat transfer coefficient is obtained from an
empirical correlation that accounts for many factors that affect the convection heat transfer. These factors, for forced and natural convection, include
the following:

112

- Geometry (diameter, surface roughness, etc.)
- Velocity
- Thermal properties
- Transport properties.

Forced and natural convection coefficients are correlated by independent formulas, and an effective combined value is obtained from the two results. The recommended equation for the combination [8] is

$$h = (h_{fc}^a + h_{ac}^a)^{1/a} ,$$

(6-13)

where

h_{fc} = the forced convection coefficient
h_{ac} = the natural convection correlation
a = an exponent that must be derived empirically.

Correlations for external and cavity receivers recommended by Siebers and Kraabel are very different. Therefore, the following discussion is divided into two sections, the first dealing with external receivers and the second dealing with cavity receivers.

6.4.1.1 External Receivers

The recommended correlation of forced convection [8] is based on the roughness of the surface. In general, the surface area of the receiver is taken as the area measured along the plane that passes through the center of the panels that absorb the heat flux. A typical receiver surface is composed of small tubes that project out of this surface. The following correlation treats these features as roughness on the surface [8]:

$k_s/D = 0.0$ (smooth surface)

(all Re_D)
$$Nu_D = 0.3 + 0.488 \, Re_D^{0.5}$$
$$[1.0 + (Re_D/282000)^{0.625}]^{0.8}$$

113

$k_s/D = 75 \times 10^{-5}$

($Re_D < 7.0 \times 10^5$: use smooth surface equation.)
($7.0 \times 10^5 < Re_D < 2.2 \times 10^7$) $Nu_D = 2.57 \times 10^{-3}\ Re_D^{0.93}$
($Re_D > 2.2 \times 10^7$) $Nu_D = 0.0455\ Re_D^{0.81}$

$k_s/D = 300 \times 10^{-5}$

($Re_D < 1.8 \times 10^5$: use smooth surface equation.)
($1.8 \times 10^5 < Re_D < 4.0 \times 10^6$) $Nu_D = 0.0135\ Re_D^{0.89}$
($Re_D > 4.0 \times 10^6$) $Nu_D = 0.0455\ Re_D^{0.81}$

$k_s/D = 900 \times 10^{-5}$

($Re_D < 1.0 \times 10^5$: use smooth surface equation.)
($Re_D > 4.0 \times 10^5$) $Nu_D = 0.0455\ Re_D^{0.81}$

where

Nu_D = the Nusselt number based on the dimension D

Re_D = the Reynolds number based on the dimension D

k_s = the surface roughness, taken to be 1/2 of the tube diameter for receiver surfaces composed of tubes

D = the characteristic dimension of the receiver

k = the conductivity of air

v = the velocity of the air stream around the receiver

ν = the kinematic viscosity for air.

In these correlations, D is intended to be the diameter of a large, cylindrical receiver. In the case of a flat-plate exposed receiver, engineering judgment is used to employ the same equations, but with the horizontal width of the receiver taken as the characteristic dimension D.

114

The natural convection coefficient is given by the expression

$$Nu_H = 0.098 \; Gr_H^{1/3} \; (T_w/T_a)^{-0.14} \; , \tag{6-14}$$

where

Nu_H = the Nusselt number based on the height of the surface H
Gr_H = the Grashof number based on the height of the surface H
β = the temperature coefficient of thermal expansion
g = the acceleration due to gravity.

This correlation does not account for surface roughness, although evidence indicates that the surface roughness effect is significant. To account for roughness, the following modification of the heat transfer coefficient given by Eq. (6-14) is recommended [8] for receiver surfaces composed of circular tubes:

$$h_{nc}(rough) = (\pi/2)h_{nc}(smooth) \; . \tag{6-15}$$

Once the natural and forced convection heat transfer coefficients are found, the effective convection coefficient is found from Eq. (6-13) using a value for the exponent a of 3.2.

For billboard-type receivers, see the measurements referred to in Ref. [4].

6.4.1.2 Cavity Receivers

Convection in cavity receivers is complicated by the presence of an aperture and the potential tilt of the cavity. In general, all surfaces within the cavity will become hot and lose heat to the atmosphere by convection. Experiments show that large convection cells form within cavity receivers, pulling ambient air into the aperture at the bottom, and hot air exits the upper part of the aperture. Air within the bulk of the cavity is replaced often enough that the ambient air within the cavity may be assumed to be the same temperature as air outside the cavity. Air that exists above the top lip of the cavity, however, will be held in place by buoyancy forces. This air will not

115

be replaced as often and can get quite hot. This has the effect of reducing the convection loss from surfaces above the top lip of the aperture. These effects are incorporated in the correlation of convection for a cavity receiver.

The basic correlation for natural convection from a cavity receiver [8] is

$$Nu_L = 0.088 \ Gr_L^{1/3} \ (T_w/T_a)^{0.18} \ , \tag{6-16}$$

where

Nu_L = the Nusselt number, based on the height of the interior of the cavity L

Gr_L = the Grashof number, based on the height of the interior of the cavity L.

Note that, since the Grashof number in Eq. (6-16) is raised to the 1/3 power, the length scale drops out of the correlation for h. An alternative to the correlation based on the properties of standard air is

$$h_{nc,0} = 0.81(T_w - T_a)^{0.426} \ . \tag{6-17}$$

This correlation must be corrected to account for the fact that air above the upper cavity lip of the cavity does not circulate. The recommended correlation [8] is

$$h_{nc} = h_{nc,0}(A_T/A_B)^{0.63} \ , \tag{6-18}$$

where

A_T = the total surface area within the cavity
A_B = the surface area within the cavity below the horizontal plane that passes through the top part of the cavity aperture.

Forced convection from a cavity receiver is not well understood. The convection heat transfer is apparently a function of wind speed and direction and

116

the geometry of the cavity. Some cavity experiments have shown wind effects; others have not.

6.4.2 Measurement of Convection Losses

Convection losses are very difficult to measure independently from other losses. One method has been used by Buna and Beninga [13]. This method exploits the variation in convection loss with temperature to isolate convection loss from that arising from other mechanisms.

An alternative method derives convection loss from measurements of total loss by subtracting losses resulting from other mechanisms. This method begins with the premise that reasonably accurate evaluations of emission, conduction, reflection, and total losses can be made. This is reasonable, because convection losses are the most difficult to measure. Convection losses are then determined by using the expression [Eq. (5-3)]

$$P_c = P_L - P_\varepsilon - P_{cd} - P_\rho \ . \tag{6-19}$$

6.4.3 Examples of Convection Losses

6.4.3.1 Calculation of Convection Losses

Table 6-2 lists conditions typical of operation of the 10-MW$_e$ Solar 1 power plant receiver [8].

Forced Convection: For these conditions, the Reynolds number is 3.17×10^6. The relative roughness for this receiver is 90×10^{-5}. This does not match the roughness value for any of the forced convection equations exactly, so interpolation will be required. Values of the Nusselt number for roughness of 75 and 300×10^{-5} are found to be 6040 and 8240, respectively. Interpolation yields a Nusselt number of 6180 for a roughness of 90×10^{-5}. This corresponds to a heat transfer coefficient of 23.0 W/m^2 °C.

117

Table 6-2. Typical Solar 1 Operating Conditions

Parameter	Value
Receiver diameter D	7.01 m
Receiver height H	13.7 m
Tube diameter d	1.27 cm
Wind velocity U_∞	7.2 m/s
Average receiver temp. T_w	400°C
Ambient temperature T_∞	20°C

Natural convection: The Grashof number for these conditions is 1.3×10^{14}. Applying Eqs. (6-14) and (6-15) yields a Nusselt number of 6930, which corresponds to a heat transfer coefficient of 13.2 W/m^2 °C.

Total convection loss: These two coefficients yield a combined coefficient of 24.2 W/m^2 °C, based on Eq. (6-13). The total convection loss is thus 2.77 MW, based on Eq. (6-12).

6.4.3.2 Measurement of Convection Losses

Table 6-3 provides measurements and calculations of thermal losses based on conditions for the advanced molten salt solar receiver tested at the CRTF in Albuquerque, N.M. [2]. In this case, the total loss was measured by using the method of complementary heliostat field partitions (see Chapter 5). The conduction losses were measured by closing the aperture door and heating the cavity electrically. The emission losses were evaluated using the CAVITY computer code [6]. The reflection losses were calculated based on absorptivity measurements. Equation (6-19) gives a value for convection loss of 150 kW.

Table 6-3. Advanced Molten Salt Receiver
Loss Evaluation

Loss Term	Value (kW)
Total loss	422
Conduction loss	12
Emission loss	184
Reflection loss	76

6.5 Reflection Losses

Reflection losses are those of solar energy from the heliostat field reflected from the receiver surface and escaping from the receiver. In general, reflection losses will be less than 15% of the total losses.

6.5.1 Calculation of Reflection Losses

In the more complex computer models used to evaluate emission losses, the reflection losses are generally taken into consideration because the absorptivity/emissivity of the absorbing and nonabsorbing surfaces are needed.

No matter how the reflection losses are evaluated, the absorptivity of the absorbing surfaces is either measured or somehow estimated. Reflection losses can then be determined from the incident power on the receiver, using

$$P_\rho = P_R(1 - \alpha) \; . \qquad\qquad (6\text{-}20)$$

6.5.2 Measurement of Reflection Losses

Absorptivity measurements are performed by using reflectometers. However, the reflectivity of the receiver is also a function of its surface characteristics; most of the receivers tested to date have a scalloped surface and this needs to be taken into account in the calculation of absorptivity and emissivity.

The HERMES system [7] uses the correlation

$$\epsilon_{eff} = \epsilon/[1 - (1 - \epsilon) \times (1 - 2/\pi)] \; , \qquad\qquad (6\text{-}21)$$

where

ϵ_{eff} = the effective emissivity

ϵ = the measured absorber emissivity,

to evaluate the effective emissivity of the scalloped surface.

119

6.5.3 Examples of Reflection Losses

An example of evaluating reflection losses is taken from Carmona [4] in the evaluation of the ASR at the Plataforma Solar. In this case, the effective emissivity is evaluated in the equation above (using measurements of the absorber emissivity). The reflective losses can be determined by either the incident power on the receiver in Eq. (6-20) or from the receiver absorbed power and the thermal losses:

$$P_\alpha = P_N + P_\epsilon + P_c + P_{cd}$$

$$P_\alpha = \alpha \, P_R \, .$$

(6-22)

Solving Eqs. (6-20), (6-21), and (6-22), we obtain

$$P_\rho = (P_N + P_\epsilon + P_c + P_{cd})(1/\alpha - 1) \, .$$

(6-23)

Other examples of evaluating reflected loss are provided by those reporting measurements at the Solar 1 power plant [14], the ASR receiver [4], and the MSS/CTE [2] in the referenced documents.

6.6 References

1. F. Kreith, Principles of Heat Transfer, Intext Press, Inc., 1973.

2. D. C. Smith and J. M. Chavez, A Final Report on the Phase I Testing of a Molten-Salt Cavity Receiver, SAND87-2290, Sandia National Laboratories, Albuquerque, NM, December 1988.

3. H. Jacobs, Receiver Losses: Results of Tests, draft report, Sandia National Laboratories, 1985.

4. R. Carmona, M. Sanchez, and H. Jacobs "Evaluation of Advanced Sodium Receiver Losses," Proceedings of the 3rd International Workshop, Konstanz, Germany, June 23-27, 1986, Solar Thermal Central Receiver Systems, Volume 1, M. Becker, ed., Cologne, Germany.

5. D. Sayers, "Cavity-A Computer Code to Couple Radiative Exchange in a Cavity Type Receiver with the Conductive-Convective Exchange to the Working Fluid," Internal Memorandum, Sandia National Laboratories, May 1985.

6. R. D. Skocypec and V. Romero, "Thermal Modeling of Solar Central Receiver Cavities," Journal of Solar Engineering, May 1989; pp. 117-123.

7. A. Brinner, Test of the Advanced Sodium Receiver (ASR) with an infrared camera system (HERMES) during test phase I of the ASR High Flux Experiment, DFVLR Internal Report IB 444 002/86, August-October 1985.

8. D. L. Siebers and J. S. Kraabel, Estimating Convective Energy Losses from Solar Central Receivers, SAND84-8717, Sandia National Laboratories, April 1984.

9. M. Abrams, The Status of Research on Convective Losses from Solar Central Receivers, SAND83-8224, Sandia National Laboratories, Albuquerque, NM, and Livermore, CA, 1983.

10. A. M. Clausing, "Convective Losses from Cavity Solar Receivers-Comparisons Between Analytical Predictions and Experimental Results," Journal of Solar Energy Engineering, February 1983.

11. R. F. Boehm, "A Review of Convective Loss Data from Solar Central Receivers," Journal of Solar Energy Engineering, May 1987.

12. A. M. Clausing, K. C. Wagner, and R. J. Skarda, "An Experimental Investigation of Combined Convection from a Short Vertical Cylinder in a Crossflow," Transactions of the ASME; Journal of Heat Transfer, Vol. 106, August 1984.

13. K. J. Beninga and T. Buna, "In Situ Evaluation of Convective Losses in Molten Salt Central Receivers," Proceedings of the ASES Meeting, 1984.

14. L. G. Radosevich, Final Report on the Power Production Phase of the 10 MW$_e$ Solar Thermal Central Receiver Pilot Plant, SAND87-8022, Sandia National Laboratories, Albuquerque, NM, March 1988.

Chapter 7

Computer Models

Computer Models

Contributed by

M. Carasso, Solar Energy Research Institute, USA

J. Guerra Macho, University of Seville, Spain

M. Kiera, INTERATOM GmbH, FRG

F. Ramos, ASINEL, Spain

M. Sánchez, CIEMAT-IER, Spain

C. Tyner, Sandia National Laboratories, USA

Contents

7.1 Introduction

Mathematical models are written to represent physical phenomena and relationships so that the characteristics of physical processes involving these phenomena can be predicted. When these processes can be represented only by complex or lengthy mathematical models, computer models are used. Many computer models have been designed to simulate different aspects of the design and operation of solar receivers. Because of the strong interaction between the solar receiver and the method used for concentrating the solar flux (e.g., a field of heliostats, a parabolic dish, a parabolic trough), many computer models simulating these concentrators have been developed that are of significance in this discussion.

It may be apparent, however, that computer models are among the least desirable ways of evaluating a solar receiver's performance, because they are the methods most removed from direct measurements of such performance. Nonetheless, as is evident in discussions in previous chapters, they are likely to play one or more important roles in the overall process used to evaluate receiver performance. For example, models can be used

• To design experiments in which either direct or indirect measurements may be taken and from which receiver performance can be calculated

• To supplement measurements whenever measurements are difficult to make (for example, convective heat transfer losses)

• To provide estimates of values for comparison with measurements.

Table 7-1 lists computer models currently used in the United States and in Europe. The models are listed according to their primary purpose. The next section of this chapter contains a brief description of each model. References are included at the end of the chapter for those interested in more detailed descriptions.

Table 7-1. Central Receiver Computer Programs Most Commonly
Used in the United States and Europe

Primary Purpose	Code Name	Originating Country	Reference
Incident Flux Estimation	DELSO1	USA	1
	MIRVAL	USA	2
	HELIOS	USA	3
	U. of Houston NS Codes	USA	4-9
Radiation Transfer in Cavity Receivers	RADSOLVER	USA	10
Energy Absorption in Working Fluid	DRAC	USA	11,12
Radiation and Energy Absorption in Cavity Receivers	CAVITY	USA	13
Radiation, Convection and Energy Absorption in Cavity Receivers	CAVITY2	USA	14
Cavity Radiation Distribution and Temperature Estimation	CAVITY/CREAM	USA	8
Volumetric Receiver Performance Evaluation	HOTAIR	USA	15
Economic Calculation for Solar Power Plant	ASPOC	Spain	16
Heat Transfer Behavior Simulation of Cavity Receiver	ATRSC	Spain	17-19
Calculations of Flux Distribution on Cavity Walls	CARE	W. Germany	20
Operational Behavior of 20-MW$_e$ GAST Plant	DYNAG	W. Germany	21,22
	GASBIE	W. Germany	23
Specific Calculations of Annual Performance of Heliostat Plants	HFLCAL	W. Germany	24
	SOLERGY	USA	25

7.2 Models Commonly Used

DELSOL

DELSOL is a performance and design optimization code that uses an analytic Hermite polynomial expansion/convolution-of-moments method (much like NS) for predicting images from heliostats. Performance is evaluated on the basis of zones that are formed by sectioning the heliostat field radially and azimuthally or on the basis of individual heliostats. Time-varying effects of insolation, cosine, shadowing and blocking, and spillage are calculated, as are time-independent effects attributable to atmospheric attenuation, mirror reflectivity, and receiver absorptivity. Although receiver radiation and convection losses are also calculated, the detail is probably insufficient for use in receiver evaluation; thus, only flux-density calculation capabilities are used for receiver analysis. Once field layout and parameters are defined, DELSOL uses much less computer time than MIRVAL or HELIOS for calculating flux-density maps.

MIRVAL

MIRVAL is a Monte Carlo ray-tracing program that simulates individual heliostats and a portion of the receiver as it calculates the optical performance of well-defined solar thermal central receiver systems. It was created for detailed evaluation and comparison of fixed-heliostat, field, and receiver designs. It accounts for the effects of shadowing, blocking, heliostat tracking, and random errors in tracking and in the conformation of the reflective surface, insolation, angular distribution of incoming solar rays to account for limb darkening and scattering, attenuation between the heliostats and the receiver, reflectivity of the mirror surface, and aiming strategy. Three receiver types (external cylinder, cylindrical cavity with a downward-facing aperture, and north-facing cavity) and four heliostat types are included in the code. It is used to calculate field efficiencies and flux maps when a rigorous optical model is required, although input and computer time requirements limit its utility in receiver evaluation.

HELIOS

HELIOS uses cone optics to evaluate flux density. It was originally developed for modeling the CRTF, where calculated flux densities can be matched to measured values. HELIOS can be used to analyze the flux density arising from fields of 1 to 559 individual heliostats or 559 cells containing multiple heliostats. Effects included in detail in HELIOS are the declination of the sun, earth orbit eccentricity, molecular and aerosol scattering in several standard clear atmospheres, atmospheric refraction, angular distribution of sunlight, reflectivity of the facet surface, shapes of focused facets, distribution of errors in the surface curvature, aiming, facet orientation, and shadowing and blocking. HELIOS is used when a detailed description of the heliostat is available and an extremely accurate evaluation of flux density is needed. As with MIRVAL, input and computer time requirements limit its utility in evaluating receivers to specific applications.

NS

The NS code, developed by the University of Houston, evaluates the optical performance of a specified central receiver field to produce flux maps for external-surround, cavity, and flat-plate types of receivers. The flux map algorithm is based on a two-dimensional Hermite expansion method. Receiver panel powers and gradients can be printed for each instant. Special timing sequences provide sunrise startup data and cloud passage data. In addition, "drift studies" provide for multiple flux maps with the heliostats either fixed (sun drift) or skewed on either axis. A typical flux calculation takes less than 30 seconds on a VAX 780.

RADSOLVER

RADSOLVER is a computer program that calculates the radiation energy transport in arbitrarily shaped solar cavity receivers. In contrast to the common assumption of gray surfaces used in the modeling of radiation transport, RADSOLVER accounts for the wavelength dependence of emission and reflection

with a band model of the radiative properties. The consideration of wavelength dependence is important in solar receiver applications where surfaces
may have significant variations over the wavelength range between solar and
thermal radiation. The phenomena included in RADSOLVER are thermal emission,
reflection and absorption of thermally emitted and solar energies, and multiple reflections of both types of radiant energy among the zones of the cavity.
Energy that would be transported within and from the cavity by convection is
not taken into account, restricting application to cavities with a stable air
mass in a windless environment. RADSOLVER calculates radiation heat transfer
among cavity zones and to working fluids, irradiation and radiosity, and temperatures on adiabatic zone surfaces.

DRAC

DRAC is not a self-contained code but rather the first in a series of driver
programs for the more general code TOPAZ. It is a relatively easy-to-use code
that permits the user to model both transient and steady-state thermohydraulic
phenomena in solar receiver tubing. Users may specify arbitrary, time-
dependent incident heat flux profiles and flow-rate changes, and DRAC will
calculate the resulting transient excursions in tube wall temperature and
fluid properties. Radiative and convective losses are accounted for. The
user may model any receiver fluid (compressible or not) for which thermodynamic data exist.

CAVITY

CAVITY is a code designed to couple the solution of the radiative exchange in
cavity-type receivers with the conduction-convection exchange to the working
fluid. Radiative losses are included in the analysis, and an estimate is made
of convective losses. However, the convective energy losses are not included
in the energy balance. The radiative calculation is performed by using a net
flux method, with a single-band analysis. The net fluxes from the radiation
energy balance are used as boundary conditions for solving the conduction-
convection governing equations. The code can be used to model any working
fluid for which property data exist. CAVITY requires that a mesh be generated

to describe the receiver geometry. The code also requires inputs of initial surface temperatures, the radiative exchange matrix, and incident solar flux for each surface. Output information includes surface temperatures and flux levels for all elements, flow rates, pressure drop, incident energy, absorbed energy, convective energy losses, radiative energy losses, and average surface temperatures.

CAVITY2

CAVITY2 is an extension of the CAVITY code. This code models the steady-state energy transfer in cavity-type receivers. The energy balance in the model includes radiative transfer between all cavity surfaces (spectrally, either a gray or two-band analysis--solar and infrared), convective transfer between the cavity surfaces and the air within the cavity, and energy transfer to the working fluid flowing in the tubes. An iterative, underrelaxed energy balance is used. Convective transfer can be modeled as either a spatially uniform, globally derived heat transfer coefficient, a spatially varying convective heat transfer coefficient, or no convective heat transfer. Inputs and outputs are similar to those of the CAVITY code.

CAVITY-CREAM

CAVITY-CREAM is a two-band (solar and IR) model for handling the thermal and radiation problem in a cavity-type receiver. It interfaces with the NS code to generate the initial solar flux distribution within the cavity. A variation of the Nusselt method is used to generate view factors between the cavity nodes. Reflected and radiated energy from each node are rescattered until absorbed or lost from the cavity aperture. Adiabatic surface temperatures are relaxed iteratively while a user-supplied model determines tube wall surface temperatures and effective IR emissivities. Special aiming routines are available to reduce peak flux levels without illuminating the aperture lip or the cavity floor or roof. The code is available from the University of Houston.

HOTAIR

HOTAIR models steady-state energy transfer in the absorber pack of a volumetric receiver. The absorber can consist of any material for which the optical and radiative properties are known and for which convective heat transfer correlations are available. The net solar and infrared radiative energy absorbed at a given depth in the absorbing pack are transferred to the working fluid by convection. The medium is assumed to be one-dimensional and homogeneous. Solar energy absorbed within the pack is modeled using a ray-trace analysis. Infrared radiative transfer is modeled using the two-flux approximation with isothermal correction, and convective energy transfer is modeled using published heat transfer correlations for wire screens. Under-relaxation is used in the iterative energy balance. Various infrared boundary conditions can be applied, and the effect of noncollimated incident flux is approximated by using effective radiative properties. Inputs to the code include solar flux, cone half-angle of incidence, inlet and outlet temperatures of the working fluid, radiative properties of the absorber pack, and a geometric description of the absorber pack. The code provides the following output: receiver thermal efficiency, working fluid mass flow rate, solar reflectance and transmittance from the absorbing pack, and infrared emittance from the absorbing pack. Fluid and matrix temperatures, solar flux absorbed, and net infrared radiative flux absorbed are all available as a function of depth within the absorbing pack.

ASPOC

ASPOC (A Solar Plant Optimization Code) was developed as part of the GAST Technological Program. Its main objective is to perform a fast economic calculation to optimize all parameters of a solar power plant. Typical applications are parametric analyses of heliostat characteristics, field lay-out, tower, receivers, thermal cycles, etc. The task consists of finding the combination of parameters that maximizes the annual thermal production per unit of the reflecting surface or minimizes the annual energy levelized cost for solar plants with cavity or cylindrical receivers, and with thermal storage or fuel support. The method consists of a stepping procedure to maximize

132

the absolute value of the desired function. It uses a selective directed search of a surrounding n-dimensional grid of points to find the increasing direction, given an initial estimation. The procedure is repeated until the improvement is small enough. The code is organized in modules to facilitate software maintenance and the updating of cost and efficiency subroutines.

ATRSC

ATRSC (Thermal Analysis of Cavity Central Receivers) calculates the heat transfer behavior of a cavity receiver. The model takes into account the radiant heat exchange among the cavity walls, the convective heat transferred between the walls and the air into the cavity, and the energy transferred to the working fluid.

The applied convective model divides the cavity into a stagnant zone and a convective zone. It calculates the heat transfer coefficients and the transferred mass between both zones. The conduction and convection heat transferred from the internal surfaces of the cavity to the working fluid are also considered, even when a fluid phase change takes place.

The input data are the geometeric description of the cavity receiver, initial temperature distribution, incoming solar flux over each surface, radiant characteristics of the cavity surfaces, and characteristics of the working fluid.

An automatic procedure discretizes the cavity receiver surfaces and calculates the view factors using Nusselt's method. The joint action of the heat transfer mechanisms is calculated by means of an iterative calculation process, which starts from the initial surface temperatures of the different elements into which the receiver has been discretized. The output data are temperatures and energy fluxes over each surface into the receiver, mass flow and temperature behavior of the working fluid, convective and radiant losses, and receiver efficiencies.

ATRSC code has been qualified with the water cavity receiver of the CESA-1 plant in Almeria, Spain. The model allows us to calculate the efficiency of a

133

given cavity receiver and to carry out a sensitivity analysis for different design parameters.

CARE

The design of cavity receivers requires the calculation of intensity distributions on the cavity walls in terms of the radiation fluxes delivered by the heliostat field in the cavity aperture. For this reason, the vectorial flux distribution of every heliostat in a specified aperture mesh has to be calculated, e.g., by applying the codes HELIOS or HFLCAL. CARE projects every flux vector onto an assembly of plant quadrilaterals forming a closed surface, including the aperture area. The user may specify the number, positions, orientations, shapes, and meshpoints of the quadrilaterals in order to simulate complex cavity geometries. By summing up the contributions incident normally on every mesh area, the program obtains the flux distribution on the walls regardless of the geometric configuration selected.

DYNAG

DYNAG is a code that simulates the non-steady-state operational behavior of the 20-MW$_e$ GAST reference CRS plant. The code simulates the open-air gas turbine cycle, including the following components:

- Compressor
- Receiver
- Pipes and mixing chambers/points
- Combustion chambers
- Gas turbine
- Waste heat boiler (air pressure losses).

The bottoming water/steam turbine cycle is not part of DYNAG simulation, because this cycle does not influence the gas turbine cycle. In addition, the dynamic process parameters that are the objects of interest of the non-steady-state analysis occur primarily in the gas cycle, while the water/steam cycle behaves more or less like a conventional one. The gas is considered to be a stationary medium, a sufficient approach for most of the operation modes in question. In addition, it saves computer time. With the subroutine REGEL, the flow characteristics of valves can be simulated for control actions.

The DYNAG code can be used primarily for non-steady-state operation modes running over the course of very short periods of time. Long-time modes (e.g., a full-day run) can be performed, too, but this consumes a lot of computer time.

The main impacts are the characteristics of the components mentioned above and the operational boundary conditions of the operation modes investigated. The following modes are investigated, including pure solar, fossil energy support, and solar/fossil hybrid operation:

- Several startup procedures
- Several clouding transients
- Shutdown
- Cooling down at night
- Load rejection.

GASBIE

The GASBIE code was designed to simulate the operation of the 20-MW$_e$ gas-cooled tower plant GAST. Because this system has no solar storage, optional use of fossil energy support is provided. For input, GASBIE requires clear insolation data, the distribution of clear and cloudy periods day by day, and heliostat field efficiencies as calculated by HFLCAL. The code determines in user-specified time steps the annual performance of the plant and the required fossil energy, depending on the load profile of the consumer. In addition, the thermal parasitics due to startup and shutdown procedures of the receiver, the fossil burner and the thermodynamic cycle are computed, whereas the electric parasitics are determined in terms of the gross electric power produced at every instant. The subsystems are specified by their partial-load curves, including dependence on relevant fluid mechanical and thermodynamical parameters under steady-state conditions. The program is structured as a constellation of routines with logical separate functions. System components, strategies of operation, etc., may be modified without fundamental structural changes to the program.

HFLCAL

The HFLCAL code, developed by Interatom, optionally calculates the annual performance of a specific configuration of heliostat field/tower/receiver/cycle and, for a fixed tower height and specified design-point conditions, the heliostat field and receiver geometry yielding the maximum annual energy per square meter of reflecting area. By repeating the optimization procedure for different tower heights and weighting the plant data with given cost functions of the plant components, the program determines the economic plant optimum. The flux reflected by a single heliostat is assumed to be a Gaussian distribution within a circular cone. The half-width is computed from the sunshape, the beam quality, the tracking error, and the astigmatic aberration on an RMS basis at each instant considered. Receiver and cycle are simulated by steady-state partial-load figures that simulate operation at a constant mass flow or at a constant temperature of the cooling media. One or two modular cavity receivers with elliptical or rectangular apertures (external flat-plate or cylindrical receivers) are optionally available. Depending on the receiver geometry, the heliostat aim is restricted to the single-point/single-line mode. The output of HFLCAL provides graphical representations of the heliostat field and of the receiver aperture/absorber in terms of isocontours of the efficiencies and of the flux distribution, respectively.

SOLERGY

SOLERGY simulates the operation of a solar central receiver power plant using an insolation record recorded at 15-minute intervals. The relatively short intervals are needed to model plant start up and the effects of cloud transients. The code has subroutines for each major plant system, i.e., heliostat field, receiver, thermal-energy storage, and turbine/generator. For each 15-minute time step, SOLERGY determines the plant's operational state (shut down, starting up, etc.) and calculates steady-state power flows through each plant system. Annual plant performance is found by summing the performance at every 15-minute time step. SOLERGY's computational algorithms are based on simple conservation of energy. There are no detailed thermodynamic calculations--no tracking of pressures and temperatures throughout the plant. In SOLERGY, user-specified plant operational parameters including the time or energy required to start up and operate a system, the parasitic load of major systems, and the performance of individual components (e.g., receiver thermal losses or turbine heat rate) are used to determine the plant's operation and performance during each time step. For an operating plant like Solar One, these data can be developed from the records of plant performance. For a plant in the design stage, fairly detailed analyses are required to develop the necessary input data.

7.3 References

1. B. L. Kistler, A User's Manual for DELSOL3: A Computer Code for Calcu-lating the Optical Performance and Optimal System Design for Solar Ther-mal Central Receiver Plants, SAND86-8018, Sandia National Laboratories, Livermore, CA, 1987.

2. P. L. Leary and J. D. Hankins, User's Guide for MIRVAL--A Computer Code for Comparing Designs of Heliostat-Receiver Optics for Central Receiver Solar Power Plants, SAND77-8280, Sandia National Laboratories, Livermore, CA, 1979.

3. C. N. Vittitoe and F. Biggs, A User's Guide to HELIOS: A Computer Code for Modeling the Optical Behavior of Reflecting Solar Concentrators, SAND81-1562 and SAND81-1180, Sandia National Laboratories, Albuquerque, NM, 1981.

4. M. D. Walzel, F. W. Lipps, and L. L. Vant-Hull, "A Solar Flux Density Calculation for a Solar Tower Concentrator Using a Two-Dimensional Hermite Function Expansion," Solar Energy, 19, pp. 239-253.

5. F. W. Lipps, Theory of Cellwise Optimization for Solar Central Receiver Systems, SAND85-8177, University of Houston Contractor Report, Sandia National Laboratories, 1985.

6. C. L. Pitman and L. L. Van-Hull, <u>Atmospheric Transmittance Model for a</u> <u>Solar Beam Propagating Between a Heliostat and a Receiver</u>, SAND83-8177, University of Houston contractor Report, Sandia national Laboratories, 1984.

7. C. L. Laurence and F. W. Lipps, <u>User's Manual for the University of</u> <u>Houston Individual Heliostat Layout and Performance Code</u>, SAND84-8187, University of Houston Contractor Report, Sandia National laboratories, 1984.

8. F. W. Lipps, "Geometric Configuration Factors for Polygonal Zones Using Nusselt's Unit Sphere," <u>Solar Energy</u>, 30 (5), 1983, pp. 413-419.

9. F. W. Lipps, <u>Generalized Layout for Collector Field with Broken Planes</u> <u>Including Modifications to the RC-Optimization CELLAY and IH-Performance</u> <u>codes</u>, Sandia Procurement 84-1637, December 1982.

10. M. Abrams, <u>RADSOLVER--A computer Program for Calculating Spectrally-</u> <u>Dependent Radiative Heat Transfer in Solar Cavity Receivers</u>, SAND81-8248, SANDIA NATIONAL LABORATORIES, LIVERMORE, CA, 1981.

11. A. F. EMERY, <u>Instruction Manual for the Program SHAPEFACTOR</u>, SAND80-8027, Sandia National Laboratories, Livermore, CA, 1980.

12. W. S. Winters, <u>DRAC--A User Friendly Computer Code for Modeling Transient</u> <u>Thermohydraulic Phenomena in Solar Receiver Tubing</u>, SAND82-8744, Sandia National Laboratories, Livermore, CA, 1983.

13. D. D. Sayers, <u>CAVITY--A Computer Code to Couple Radiative Exchange in a</u> <u>Cavity-Type Receiver with the Conductive-Convective Exchange to the</u> <u>Working Fluid</u>, Internal Memorandum RS8245/20, Sandia National Laboratories, Livermore, CA.

14. J. M. Chavez, <u>Molten Salt Subsystem/Component Experiment; Receiver Sub-</u><u>system Test</u>, Final Report, SAND87-2290, Sandia National Laboratories, in preparation.

15. R. D. Skocypec, <u>Heat Transfer Analysis of the IEA Volumetric Receiver</u>, SAND87-2969, Sandia National Laboratories, in preparation.

16. F. Ramos, J. Mateos, and J. de Marcos, "Optimization of a Central Receiver Solar Electric Power Plant by the ASPOC Program," <u>Fourth</u> <u>International Symposium on Research, Development and Applications of</u> <u>Solar Thermal Technology</u>, June 13-17, 1988, Santa Fe, NM, USA.

17. J. Guerra, J. Coronado, and S. Alvarez, "Modelizacion de la Transferencia Convectiva-Radiante en Receptores Solares de Cavidad," <u>Energia</u>, 1987, pp. 123-128.

18. J. Coronado, J. C. Cuadrado, J. Guerra, and R. Velazquez, "A Simulation Model for Solar Central Receivers. Application to CESA-1," <u>XXIII</u> <u>Congreso de la Cooperacion Mediterranea para la Energia Solar</u>, V. Ruiz, y M. Garcia, eds., Publicaciones ADESA, Sevilla, Spain, 1985.

19. J. Guerra, S. Alvarez, and J. Coronado, "Modelizacion del Intercambio Termico en Receptores Solares de Cavidad," <u>2° Congresso Iberico de la</u> <u>Energia Solar</u>, ISES, Lisbon, Portugal, 1984.

20. W. Pinterowitsch, <u>CARE--a code for calculating the flux density on the</u> <u>inner walls of a cavity receiver</u> (in German), GAST-IAS-BT-200000-008, Interatom, 1982.

21. K.-H. Jansen, <u>DYNAG Code for Investigation of Non-Steady-State Operation</u> <u>Modes of the GAST Reference Plant</u> (in German), IAS-BT100200-063, Interatom GmbH, GAST Technology Program, 06.10.83.

22. J. Guillen, <u>System Dynamic Behaviour</u>, Asinel/Spain, GAST Technology Program, Final Presentation in Lahnstein/FTH, 30./31.05.88.

23. M. Fsadni, <u>Description of the program GASBIE for the simulation of the GAST plant operation</u>, GAST-IAS-BT-100200-057, Interatom, 1983.

24. M. Kiera, <u>Description of the program system HFLCAL</u> (in German), GAST-IAS-BT200000-075, Interatom, 1986.

25. D. J. Alpert and G. J. Kolb, <u>Performance of the Solar One Power Plant as Simulated by the SOLERGY Computer Code</u>, SAND88-0321, Sandia National Laboratories, Albuquerque, NM, April 1988.

Solar Thermal Energy Utilization

*German Studies in Technology
and Application*

Volume 1: **M. Becker** (Ed.)

General Investigations on Energy Availability

1987. Softcover DM 85,– ISBN 3-540-18028-1

Volume 2: **M. Becker** (Ed.)

Technologies of Heat Exchangers and Storage

1987. Softcover DM 85,– ISBN 3-540-18031-1

Volume 3: **M. Becker** (Ed.)

Solar Thermal Energy for Chemical Processes

1987. Softcover DM 170,–
ISBN 3-540-18032-X

1987. Volumes 1–3 (as a set).
Softcover DM 295,–
ISBN 3-540-18033-8

Springer-Verlag
Berlin
Heidelberg
New York
London
Paris
Tokyo
Hong Kong
Barcelona

C.-J. Winter, DLR Stuttgart; **R. Sizmann,** Universität
München; **L. Vant-Hull,** University of Houston, Eds.)

Solar Power Plants

Fundamentals – Technology – Systems – Economics

1991. Approx. 350 pp. Approx. 150 figs. Hardcover. In prep.
ISBN 3-540-18897-5

Solar Thermal Central Receiver Systems

Volume 1 and 2: **M. Becker** (Ed.)

Proceedings of the Third International Workshop,
June 23–27, 1986, Konstanz, Federal Republic of Germany

1986. (Not available separately) Softcover DM 148,–
ISBN 3-540-17062-9

M. Becker, M. Böhmer (Eds.)

GAST – The Gas-Cooled Solar Tower Technology Program

Proceedings of the Final Presentation

1988. Softcover DM 148,–
ISBN 3-540-50121-5

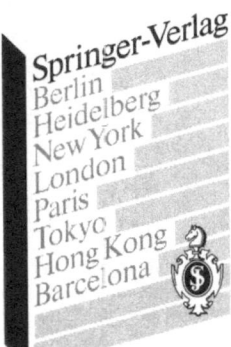

Springer-Verlag
Berlin
Heidelberg
New York
London
Paris
Tokyo
Hong Kong
Barcelona